*FM 4-0

Field Manual
No. 4-0

Headquarters
Department of the Army
Washington, D.C., 30 April 2009

Combat Sustainment

Contents

Distribution Restriction: Approved for public release; distribution unlimited.

***This publication supersedes FM 4-0, 29 August 2003.**

AF797323

Figures

Tables

Preface

FM 4-0 is the Army's keystone manual for sustainment. Its purpose is to provide authoritative doctrine for the sustainment of forces in full spectrum operations. It expands on sustainment doctrine introduced in FM 3-0 to incorporate a broader approach to sustaining the fighting force. The intent of this FM is to describe how sustainment builds and maintains combat power, supports strategic and operational reach, and enables endurance. This FM establishes how sustainment operations are integrated and synchronized into the overall operations process – plan, prepare, execute, and assess.

In accordance with FM 3-0, the terms combat arms, combat support (CS), and combat service support (CSS) are no longer used as Army doctrine lexicon. The term sustainment covers many of the functions, tasks, and organizations formerly described under CSS although there are some differences which will be discussed in this FM. Sustainment includes the major sub-functions logistics, personnel services, and health services support. Sustainment as defined and discussed in this FM is compatible with the joint function, sustainment.

FM 4-0 has five chapters and two appendices. It addresses the operational environment and modular force sustainment organizations. It introduces the roles and functions of the generating and operating forces. It explains the art of battle command from a sustainment commander's perspective. Finally, it explains sustainment as one of six warfighting functions (WFF) and presents doctrine for how sustainment is planned, prepared, executed, and continuously assessed.

- Chapter 1 is the introduction to sustainment. It covers the principles of sustainment and the sustainment WFF.
- Chapter 2 covers the roles and responsibilities of organizations providing sustainment in support of unified action. It links Army sustainment to joint sustainment operations. It covers commands, organizations, and agencies that provide sustainment including Joint organizations, Army Generating Forces, and Operating Forces. It also discusses sustainment to multinational operations.
- Chapter 3 covers command and control. It discusses battle command from a sustainment perspective. It also discusses modular force sustainment organization and their relationships to Army operations.
- Chapter 4 discusses the integration of sustainment into operations. It explains how sustainment considerations are integrated into the operations process.
- Chapter 5 provides a more detailed discussion on the major functional elements of sustainment.
- Appendix A contains details on the various information systems.
- Appendix B lists the references for multinational Standardization Agreements (STANAGs) impacting on the sustainment of forces.

FM 4-0 applies to the Active Army, the Army National Guard/Army National Guard of the United States, the United States Army Reserve unless otherwise stated, and military and civilian leaders at all levels of war. It provides guidance to our joint and multinational partners on how Army sustainment may support joint and multinational operations. It links Army sustainment doctrine to joint sustainment doctrine as expressed in joint doctrinal publications, specifically, JP 1-0 series and JP 4-0 series manuals.

FM 4-0 uses joint terms where applicable. Most terms with joint or Army definitions are in the glossary and the FM. Glossary references: Terms for which FM 4-0 is the proponent publication (the authority) have an asterisk in the glossary. Text references: Definitions for which FM 4-0 is the proponent publication are in boldfaced text. These terms and their definitions will be in the next revision of FM 1-02. For other definitions in the FM, the term is italicized and the number of the proponent publication follows the definition. Headquarters, U.S. Army Training and Doctrine Command (HQUSATRADOC), is the proponent for this publication. The preparing agency is the Combined Arms Support Command (CASCOM), Concepts and Doctrine Directorate. Send written comments and recommendations on a DA Form 2028 (Recommended Changes to Publications and Blank Forms) to Commander, U.S. Combined Arms Support Command, ATTN: ATCL-CDC-DJ (FM 4-0), 2221 Adams Ave, Fort Lee, VA 23801- 2102 or by e-mail to Concepts&DoctrineDirPersonnel@conus.army.mil; or submit an electronic DA Form 2028.

INTRODUCTION

FM 4-0, *Combat Service Support*, was published under the new Army doctrine numbering system on August 2003. The manual served two purposes. First, it linked CSS and operational doctrine. Second, it served as the bridge between joint doctrine and Army CSS doctrine. Since its publication in 2003, the world and the Army have changed.

The new FM 4-0, *Sustainment,* implements the changes to our doctrine as a result of the conversion to a modular Army. It discusses from a broad perspective how sustainment is provided to the modular Army while conducting full spectrum operations. It also serves as the bridge between joint and Army sustainment doctrine.

THE SUSTAINMENT OPERATIONAL CONCEPT

The Army sustainment operational concept is the provision of logistics, personnel services, and health service support necessary to maintain and prolong operations until successful mission completion. This is accomplished through the integration of national and global resources and ensures Army forces are physically available and properly equipped, at the right place and time, to support the combatant commander (CCDR). The concept leverages host nation (HN) and multinational support, contracting, and other available capabilities to reduce over burdening military resources and at the same time maintaining a campaign quality Army.

The Army sustainment operational concept is based on an integrated process (people, systems, materiel, health services, and other support) inextricably linking sustainment to operations. The concept focuses on building a combat ready Army force, delivering it to the CCDR as part of the joint force, and sustaining its combat power across the depth of the operational area and with unrelenting endurance. This is accomplished through generating forces, consisting of Army organizations whose primary mission is to generate and sustain the operational Army.

It is the purposeful reliance by one Service's forces on another Service's capabilities to maximize the complementary and reinforcing effects of both. Through joint sustainment interdependence, Army forces have strategic and operational reach supported by a continuous flow of sustainment enabling CCDR's freedom of action to achieve national military strategy and decisive victory.

The quality of force readiness is measured by its sustainment. Sustainment builds Army forces by manning it with trained Soldiers and leaders; equipping it with the materiel (individual and unit); maintaining Soldier and Family readiness; and sustaining it for full spectrum operations. This is enabled by an integrated network of information systems linking sustainment to operations. As a result, commanders at all levels see the operational environment understand what is needed, track what is requested, and make crucial decisions ensuring responsive sustainment.

Much has changed in the Army since September 11, 2001. The Army has transitioned from a forward deployed force to a CONUS based force, from a division centric structure to a brigade centric structure. We are fighting two simultaneous wars while concurrently modernizing our force. Modularity has been a significant transformation for the Army. The remainder of the Introduction will highlight some of the sustainment changes in doctrine and force structure.

COMBAT SERVICE SUPPORT TO SUSTAINMENT

One of the most notable changes is the title of the FM from Combat Service Support to Sustainment. The Army made a conscious decision to rescind the terms combat arms, combat support, and combat service support. It now uses the appropriate WFF to describe unit types and functions. The six WFFs replace the battlefield operating systems and consist of movement and maneuver, intelligence, fires, sustainment, command and control (C2), and protection. The six WFFs also make up the elements of combat power tied together by leadership and information (see Figure 1). The sustainment WFF describes both unit types and functions.

The Sustainment WFF consists of three major sub-functions—logistics, personnel services, and health services. It should be noted that there is a realignment of some of the former CSS BOS tasks across several Army WFFs and vice versa. Some of these realignments are as follows.

- Explosive ordnance disposal (EOD). EOD tasks are aligned with the protection WFF. Proponency for EOD resides with the Ordnance Center and School and the tasks are performed by EOD forces (see FM 4-30.50).
- Force health protection (FHP) tasks such as preventive dentistry and preventive medicine tasks also fall under the protection WFF. Proponency for these tasks resides with the Army Medical Department Center and School. These tasks are performed by trained medical personnel (see FM 4-02).

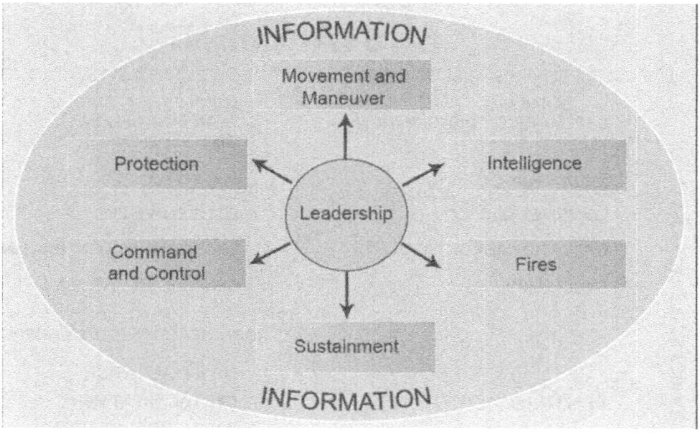

Figure 1. Warfighting Functions

JOINT SUSTAINMENT AND THE ARMY SUSTAINMENT WARFIGHTING FUNCTION

FM 4-0 serves as a bridge between joint doctrine and Army doctrine. As such, it is important to understand the linkage between sustainment as a joint function and sustainment as an Army WFF.

The joint function sustainment is one of six joint functions; C2, intelligence, fires, movement and maneuver, protection, and sustainment. Joint functions are related capabilities and activities grouped together to help Joint Force Commanders (JFC) integrate, synchronize, and direct joint operations. The joint functions are mirrored by the Army WFFs. Joint Pub 3-0 describes joint sustainment as "the provision of logistics and personnel services necessary to maintain and prolong operations until mission accomplishment". Sustainment is primarily the responsibility of the supported CCDR and Service component commanders in close cooperation with the Services, Chief of Staff of the Army (CSA), and supporting commands (JP 4-0).

The Army sustainment WFF is one of six WFFs. A WFF is a group of tasks and systems (people, organizations, information, and processes) united by a common purpose that commanders use to accomplish missions and training objectives (FM 3-0). The Army sustainment WFF is fundamentally linked to the joint function sustainment, but there are subtle differences, as demonstrated in Figure 2.

First it should be noted that the joint functions are capability based. The JFC must rely on the Services to provide the capabilities upon which he/she conducts joint operations. On the other hand, Army forces own their functional capabilities and generally describe those capabilities as functions or tasks. Secondly, the joint function sustainment is split between two staff elements, the Joint Staff (J-1), Personnel and Joint Staff (J-4), Logistics. Doctrine for these two staff elements is found in JP 1-0, Joint Personnel Services and JP 4-0, Joint Logistics.

Another point is the joint function sustainment consists of two major sub functions—logistics and personnel services. The Army Sustainment WFF consists of three major sub functions—logistics, personnel services, and health service support (HSS). In the joint function sustainment health services is a sub function of logistics. Figure 2 crosswalks the sub functions between joint and Army sustainment.

SUSTAINMENT

JOINT SUSTAINMENT FUNCTION				ARMY SUSTAINMENT WARFIGHTING FUNCTION
JOINT LOGISTICS CAPABILITIES	SUPPLY DEPLOYMENT / DISTRIBUTION	SUPPLY TRANSPORTATION DISTRIBUTION	LOGISTICS	
	MAINTENANCE	MAINTENANCE		
	LOGISTICS SERVICES	FIELD SERVICES		
	OPERATIONAL CONTRACTING	OPERATIONAL CONTRACTING		
	ENGINEERING	GENERAL ENGINEERING		
	HEALTH SERVICES	ARMY HEALTH SYSTEMS SUPPORT HOSPITALIZATION DENTAL TREATMENT BEHAVIORAL HEALTH LABORATORY SERVICES CBRNE TREATMENT MEDICAL EVACUATION MEDICAL LOGISTICS	ARMY HEALTH SERVICES	
PERSONNEL SERVICES	PERSONNEL LEGAL CHAPLAIN FINANCE	HUMAN RESOURCES SUPPORT LEGAL SUPPORT RELIGIOUS SUPPORT FINANCIAL MANAGEMENT BAND SUPPORT	PERSONNEL SERVICES	

Figure 2. Sustainment Crosswalk

THE OPERATIONAL ENVIRONMENT

The operational environment is defined as the composite of the conditions, circumstances, and influences that affect the employment of capabilities and bear on the decisions of the commander (FM 3-0). The operational environment directly impacts the means by which the Army sustains operations as a part of the joint and multinational force (MNF). It is interconnected and increasingly global. It is also extremely fluid, with continually changing coalitions, alliances, partnerships, and new threats constantly appearing and disappearing. The constantly changing operational environment presents many challenges to sustainment of forces. These challenges include providing support in varied physical environments (terrain, climate, and urban areas), working among multicultural populations and operating in areas where it may be difficult to discern the enemy from non combatants. In today's global situations, the physical environment alone in one area of responsibility (AOR) may differ vastly from another. The biggest challenge may be providing responsive sustainment to a rapidly deployable Army force to meet threats worldwide. Overcoming this challenge requires more than ever, joint sustainment interdependence. All Services require logistics, personnel services, and HSS to maintain operational readiness and combat effectiveness. Working together as a joint team enables the U.S. military to reduce redundancies and increase efficiencies in sustainment operations.

Therefore, the operational environment requires forging strong sustainment alliances and coalitions. It will be rare that U.S. forces operate alone. As such, sustainment commanders must consider and plan for interoperability with our allies and coalition forces. Commanders may be required to share technology, processes, and procedures to ensure that our partners can deliver the same decisive operations. Commanders should also consider contracting and host nation support (HNS) options as possible sources of support. Contracting provides commercial supplies, services, and minor construction to supplement military capabilities, giving the mission commander operational flexibility. HNS provides trained, skilled labor to augment sustainment operations.

The operational environment includes the spectrum of conflict which ranges from peace to general war. We will continue to see natural or man-made disasters. As a result, the U.S. military will find itself providing stability to weakened, failed, or defeated governments. Sustainment provides the necessities of life. During Stability operations, sustainment may be critical to influencing military strategy and gaining support of affected populations. While sustainment may not be the decisive operation, it can serve to shape the environment. Sustainment commanders must understand how to maximize benefits afforded by working with nongovernmental organizations (NGO), other governmental agencies (OGA), and intergovernmental organizations (IGO).

We face a variety of threats to our homeland ranging from natural disasters to direct attacks. The employment of our military to assist in responding to these threats is a reality. As a result, the Army reserves and active components must plan for and be prepared to provide the required sustainment resources.

THE MODULAR FORCE

Numerous operations conducted over the past two decades have demonstrated that Army of Excellence organizations were not as flexible and responsive as Joint Force Commanders (JFCs) required. They met JFC needs, but at high costs in organizational turbulence, inefficiency, and slower response times than desired.

The Army modular organizations provide a mix of land combat power capabilities that can be organized for any combination of offensive, defensive, stability, or civil support operations as part of a joint campaign. The modular force has brought about many changes to the Army's capability to provide sustainment. These changes cut across all of the sustainment functions and represent the Army's imperatives.

- A modular "brigade-based" Army that is more responsive to geographic combatant commanders' (GCC) needs, better employs joint capabilities, facilitates force packaging and rapid deployment, and fights as self-contained units in non-linear, non-contiguous operational environment.
- An Army sustainment structure that is responsive to the needs of a Joint and Expeditionary campaign quality Army.
- Eliminates redundancy and streamlines support by reducing unnecessary layers.
- Provides a sustainment capability that leverages emerging technologies, links support to supported organizations, and the Army to Joint organizations – from continental United States (CONUS) to area of operations (AO) and within area of responsibilities (AORs).

FM 4-0 will discuss these changes in greater detail in the coming chapters but a few highlighted changes are emphasized below.

ARMY FORCE GENERATION (ARFORGEN)

ARFORGEN is the Army's system for generating land power capabilities that respond to the operational needs of JFCs and sustaining those capabilities as long as required. It is a shift from tiered readiness to cyclic readiness and represents a change of the way the Army generating force performs its Title 10 functions. The generating force resets operational forces upon redeployment. A more detailed discussion of generating force roles and responsibilities is found in Chapter 2.

The overarching purpose of ARFORGEN is to provide CCDRs and civil authorities with trained and ready units. Under ARFORGEN, the US Army Materiel Command (USAMC) is responsible, in conjunction with the Assistant Secretary for Acquisition (ASA) Acquisition, Logistics, and Technology (ALT), for resetting the force and providing acquisition, logistic, technology, and contingency contracting support to globally deployed operational forces. Reset refers to the systematic restoration of deployed units to an appropriate level of equipment, Soldier, and Family readiness in preparation for future deployments and contingencies. Selected support activities, within an installation's Directorate of Logistics from the Installation Management Command (IMCOM) assist in resetting equipment and generating forces as required.

DISTRIBUTION AND MATERIEL MANAGEMENT

Modularity brought changes to how distribution and materiel management are performed, especially at echelons above brigade. Distribution and materiel management are combined under the distribution management centers (DMC) of the theater sustainment command (TSC) and sustainment brigade (Sust Bde). Unlike the Army of Excellence structure, this has streamlined the management of logistics support to operations making it a more capable and efficient support structure.

Medical logistics material management is performed by the Medical Command (Deployment Support) (MEDCOM (DS)), Medical Logistics Management Center (MLMC) forward support team, and medical logistics companies in the Multifunctional Medical Battalion (MMB).

Distribution management is enabled by automation capabilities such as the Single Army Logistics Enterprise (SALE) comprised of the enterprise elements of the Logistics Modernization Program, Global Combat Service Support System-Army Field Tactical and the Army Enterprise System Integration Program. The enterprise nature of the SALE with its improved supporting processes, business intelligence, and decision support tools coupled with the situational awareness of the Battle Command Sustainment Support System (BCS3) will provide improved support to the supported commander's requirements. A more detailed discussion of distribution management is in Chapter 4.

HUMAN RESOURCES SUPPORT (HRS)

Modular HRS organizations provide the commander with tailorable, flexible options to cope with the many challenges encountered during military operations. Personnel services delivery redesign (PSDR) makes possible independent Human Resource (HR) operations, regardless of location. The redesign leverages web based connectivity and bandwidth to support an expeditionary Army. A combination of increased training, new equipment, more robust communications and additional manpower, has made possible S-1-centric HR support to the Brigade Combat Team (BCT). S-1 centric HR support directly links the S-1 with the US Army Human Resources Command and enables the S-1 to execute all Soldier essential personnel services (EPS) (promotions, ID cards, actions, and so on), functions and capabilities at brigade level. At the theater level, the Human Resources Sustainment Center (HRSC), HR companies, platoons, and teams execute and support all HR operations performing casualty, postal, and personnel accounting in the theater.

FINANCIAL MANAGEMENT (FM) OPERATIONS

FM operations have transformed to better support the Army by integrating finance and resource management (RM) capabilities. These capabilities are executed by properly sized, modular FM structures that provide the ability to deploy the right mix of FM units based on Mission, Enemy, Terrain and

Weather, Troops and Support Available, Time Available and Civil Considerations (METT-TC). Within the TSC, the Financial Management Center (FMC), FM companies and FM detachments execute finance operations including disbursing, accounting, contracting support, and theater cash management. RM capability is found at the Theater Army, TSC, Expeditionary Sustainment Command (ESC), Division, and Corps.

ARMY HEALTH SYSTEM SUPPORT

The Army Medical Department (AMEDD) is transforming to a modular force with smaller deployable units. The AMEDD modular force, as exemplified by new unit designs, is versatile, scalable, and possesses standardized medical capability packages that can be easily deployed in support of full spectrum operations. Among those new unit designs are the medical command (deployment support) (MEDCOM [DS]) that serves as the theater medical force provider, the medical brigade (MEDBDE), which consists of early entry and expansion modules tailorable to meet specific mission requirements, and the MMB that replaces area support, medical logistics, and evacuation battalions. This new force structure emphasizes "early entry" and improved tactical mobility operational capabilities. These and other modularization initiatives result in the deployed medical force occupying less space in the operational environment and placing fewer demands on transportation in terms of weight and space.

Chapter 1

Introduction to Sustainment

Sustainment is the provision of logistics, personnel services, and HSS necessary to maintain operations until mission accomplishment (FM 3-0). The Sustainment operational concept supports the Army's operational concept of full spectrum operations as described in FM 1 and FM 3-0. The provision of sustainment is an integrated process (people, systems, materiel, health services, and other support) inextricably linked to operations. From a strategic perspective sustainment builds Army combat readiness, delivers a combat ready Army to the CCDR as part of the joint force, and maintains combat power and endurance across the depth of the operational area. This is supported by Army generating forces whose unbreakable link enhances Army forces reach. It is joint interdependent, relying on and providing to other Services with capabilities to support the CCDRs goals. At the operational and tactical levels, sustainment is provided by highly trained modular sustainment forces, integrated and synchronized with the operational plan. They are supported by automated systems that precisely track requirements which give commanders the time and information to make informed support decisions. They support commander's needs to provide committed forces with flexible support for their operations.

This FM will discuss how sustainment supports full spectrum operations. This chapter will begin by laying out the basics with a discussion of the principles of sustainment and the functional elements of the sustainment WFF.

PRINCIPLES OF SUSTAINMENT

1-1. The principles of sustainment are essential to maintaining combat power, enabling strategic and operational reach, and providing Army forces with endurance. The principles are integration, anticipation, responsiveness, simplicity, economy, survivability, continuity, and improvisation.

1-2. While these principles are independent, they are also interrelated (see Figure 1-1). For example, in order for commanders to provide responsive sustainment, they must be able to anticipate requirements based on their knowledge and understanding of future operations. Simplicity in planning and executing sustainment increases survivability, improves efficiencies through economy, and facilitates a continuity of resources thus reducing complexity and confusion. When the execution of plans does not proceed as expected, commanders may improvise to meet mission requirements. The most essential principle is integration. Without deliberate integration of Army sustainment with Joint and MNFs and OGA the achievement of these principles becomes impossible.

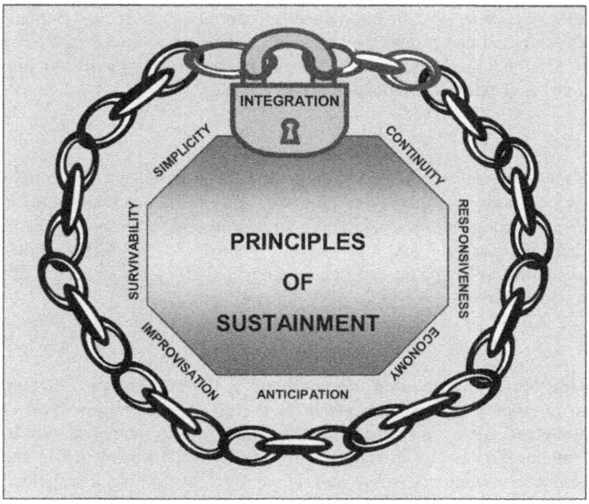

Figure 1-1. Principles of Sustainment

INTEGRATION

1-3. Integration is the most critical principle. **Integration is joining all of the elements of sustainment (tasks, functions, systems, processes, and organizations) to operations assuring unity of purpose and effort.** It requires deliberate coordination and synchronization of sustainment with operations across all levels of war. Army forces integrate sustainment with joint forces and multinational operations to maximize the complementary and reinforcing effects from each service component's or nation's competencies and resources. Integration of sustainment occurs throughout the operations process—plan, prepare, execute, and assess. One of the primary functions of the sustainment staff is to ensure the integration of sustainment with operations plans. Not properly integrating sustainment and operations could result in mission failure.

ANTICIPATION

1-4. Anticipation is the ability to foresee events and requirements and initiate necessary actions that most appropriately satisfy a response. Anticipation of sustainment facilitates responsive support. It is based on professional judgment resulting from experience, knowledge, education, intelligence, and intuition. Sustainment commanders and staffs visualize future operations and identify appropriate required support. They must then start the process of acquiring the materiel or placement of support that bests sustains the operation. Anticipation is facilitated by automation systems that provide the common operational picture upon which judgments and decisions are based. Anticipating sustainment also means staying abreast of operational plans (OPLANs), continuously assessing requirements, and tailoring support to meet current operations and the changing operational environment.

RESPONSIVENESS

1-5. Responsiveness is the ability to meet changing requirements on short notice and to rapidly sustain efforts to meet changing circumstances over time. It is providing the right support in the right place at the right time. It includes the ability to see and forecast operational requirements. Employing appropriate information systems enables the commander to make rapid decisions. Responsiveness involves identifying, accumulating, and maintaining sufficient resources, capabilities, and information necessary to meet rapidly changing requirements. A responsive sustainment system is crucial to maintaining endurance; it provides the commander with flexibility and freedom of action. It also maintains the tempo of operations and the ability to retain and exploit the initiative. Through responsive sustainment, commanders maintain operational focus and pressure, set the tempo of friendly operations to prevent exhaustion, replace ineffective units, and extend operational reach.

SIMPLICITY

1-6. Simplicity strives to minimize the complexity of sustainment. Simplicity relates to processes and procedures. Unnecessary complexity of processes and procedures compounds the confusion. Simplicity fosters efficiency throughout the operations process and allows for more effective control of sustainment. Clarity of tasks, standardized and interoperable procedures, and clearly defined command relationships contribute to simplicity. Simplicity enables economy and efficiency in the use of resources, while ensuring effective support of forces.

ECONOMY

1-7. Economy means providing sustainment resources in an efficient manner to enable a commander to employ all assets to generate the greatest effect possible. The commander achieves economy through efficient management and discipline by prioritizing and allocating resources. Staffs look for ways to eliminate redundancies and capitalize on joint interdependencies. They also apply discipline in managing resources minimizing waste and unnecessary stockpiling. Disciplined sustainment assures the greatest possible tactical endurance of the force and constitutes an advantage to commanders who achieve economy of force in sustainment. Staffs also achieve economy by contracting for support or using HN resources that reduce or eliminate the use of limited military resources. Economy reflects the reality of resource shortfalls, while recognizing the inevitable friction and uncertainty of military operations. Economy enables strategic and operational reach by reducing unnecessary use of transportation requirements. Additionally, it reduces unnecessary storage and warehouse support.

SURVIVABILITY

1-8. Survivability is the ability to protect personnel, information, infrastructure, and assets from destruction or degradation. It includes all aspects of protecting personnel (includes FHP), materiel, and organizations while deceiving the enemy. The ability of adversaries to disrupt the flow of sustainment could significantly degrade forces' ability to conduct operations as well as sustain them. Planners integrate survivability with operational planning to maximize survivability. Dispersion and decentralization of sustainment functions enhances survivability. The commander may have to balance risk with survivability in considering redundant capabilities and alternative support plans. The ability to protect lines of communications promotes survivability, helping to ensure operational reach and endurance.

CONTINUITY

1-9. Continuity is the uninterrupted provision of sustainment across all levels of war. Continuity is achieved through a system of integrated and focused networks linking sustainment to operations. Continuity is enabled through joint interdependence, linked organizations, distribution systems, and information systems. Continuity assures confidence in sustainment allowing commanders freedom of action, operational reach, and endurance. It requires commanders to track resources and make critical decisions eliminating backlogs or bottlenecks. Sustainment staffs at all levels work hand in hand with operational staffs ensuring synchronization of requirements over the entire course of the operation.

IMPROVISATION

1-10. Improvisation is the ability to adapt sustainment operations to unexpected situations or circumstances affecting a mission. It includes creating, inventing, arranging, or fabricating what is needed from what is available. It may also involve changing or creating methods that adapt to an enemy that quickly evolves. This requires commanders, their staffs, and Soldiers to improvise other possible means to accomplish an operation. The sustainment commander must apply operational art to visualize complex operations and understand what is possible at the tactical level. These skills enable commanders to improvise operational and tactical actions when enemy actions or unexpected events disrupt sustainment operations.

THE SUSTAINMENT WARFIGHTING FUNCTION

1-11. *The sustainment WFF is related tasks and systems that provide support and services to ensure freedom of action, extend operational reach, and prolong endurance (FM 3-0).* The endurance of Army forces is primarily a function of their sustainment. Sustainment determines the depth and duration of Army operations. Successful sustainment enables freedom of action by increasing the number and quality of options available to the commander. It is essential to retaining and exploiting the initiative. The sustainment WFF consists of three major sub-functions: logistics, personnel services, and health services support. A summary of the categories and functions of sustainment is outlined below. Chapter 5 provides a detailed discussion of the elements of the sustainment WFF.

1-12. FM 3-0 places Internment/Resettlement (I/R) operations under the sustainment WFF. *Internment/Resettlement operations take or keep selected individuals in custody or control as a result of military operations to control their movement, restrict their activity, provide safety, and/or gain intelligence (FM 3-19.40).* I/R operations comprise those measures necessary to guard, protect, sustain, and account for people that are captured, detained, confined, or evacuated from their homes by the U.S. armed forces.

LOGISTICS

1-13. *Logistics is the planning and executing the movement and support of forces. It includes those aspects of military operations that deal with: design and development, acquisition, storage, movement, distribution, maintenance, evacuation, and disposition of materiel; movement, evacuation, and hospitalization of personnel; acquisition or construction, maintenance, operation, and disposition of facilities; and acquisition or furnishing of services (JP 4-0).* As noted in Figure 2, there are some differences from joint and Army logistics. For example Army HSS is not considered a logistics function.

SUPPLY

1-14. Supply is the procurement, distribution, maintenance while in storage, and salvage of supplies, including the determination of kind and quantity of supplies. There are ten classes in the U.S. supply system: CL I Subsistence; CL II Clothing and Individual Equipment; CL III, Petroleum and Solid Fuels; CL IV, Construction Materiel; CL V, Ammunition; CL VI, Personal Demand Items; CL VII, Major end Items, CL VIII, Medical Materiel; CL IX, Repair Parts and Components; and CL X, Non Military Materiel. While munitions is a class of supply, it is unique due to the complexities of activities associated with its handling. Munitions are devices charged with explosives, propellants, pyrotechnics, initiating composition, or nuclear, biological, or chemical material for use in military operations, including demolitions. Munitions require special shipping and handling, storage, accountability, surveillance, and security. Munitions operations include the critical functions of maintenance, repair, recovery, demilitarization, as well as initial control and management of captured enemy ammunition.

FIELD SERVICES

1-15. Field services maintain combat strength of the force by providing for its basic needs and promoting its health, welfare, morale, and endurance. They include clothing repair and exchange, laundry and shower support, mortuary affairs (MA), aerial delivery, food services, billeting, and sanitation. All field services receive the same basic Army-wide priority, but the commander decides which are most important. FM 4-20.1 provides a full discussion on field services.

MAINTENANCE

1-16. Maintenance is all actions taken to retain materiel in a serviceable condition or to restore it to serviceability. It consists of two levels, field and sustainment maintenance. It includes inspection, testing, servicing, and classification as to serviceability, repair, rebuilding, recapitalization, reset, and reclamation. It also includes all supply and repair actions taken to keep a force in condition to carry out its mission.

TRANSPORTATION

1-17. Transportation is the moving and transferring of personnel, equipment, and supplies to support the concept of operations, including the associated planning, requesting, and monitoring. Transportation plays a key role in facilitating deployment and distribution. Transportation includes military, commercial, and multinational capabilities. Transportation assets include surface and air modes, terminal and movement control units, activities, and infrastructure. See Chapter 5 for more detail and FM 55-1 for a discussion of transportation operations.

DISTRIBUTION

1-18. Distribution is defined as the operational process of synchronizing all elements of the logistics system to deliver the right things to the right place and right time to support the CCDR. It is a diverse process incorporating distribution management and asset visibility. See Chapter 4 for more detail and FM 4-01.4.

OPERATIONAL CONTRACT SUPPORT

1-19. Operational contract support is the process of planning for and obtaining supplies, services, and construction from commercial sources in support of operations along with the associated contractor management functions. Deployed U.S. forces rely increasingly on contracting to supplement organic sustainment capabilities and on contractors to perform a growing percentage of many sustainment functions. JP 4-10 and FMI 4-93.42 cover the roles and responsibilities for planning and managing contracting in support of operational commanders.

GENERAL ENGINEERING SUPPORT

1-20. *General engineering includes those engineering capabilities and activities, other than combat engineering, that modify, maintain, or protect the physical environment. Examples include: the construction, repair, maintenance, and operation of infrastructure, facilities, lines of communication and bases, and terrain modification and repair and selected explosive hazard activities (JP 3-34).* Engineering provides construction support, real estate planning and acquisition, and real property maintenance responsive to environmental considerations. General engineering support is discussed further in Chapter 5. See FM 3-34 and FM 3-34.400 for more information.

PERSONNEL SERVICES

1-21. Personnel services include HR support, religious support, FM, legal support, and band support. **Personnel services are those sustainment functions maintaining Soldier and Family readiness and fighting qualities of the Army force.** Personnel services complement logistics by planning for and coordinating efforts that provide and sustain personnel (FM 3-0).

HUMAN RESOURCE (HR) SUPPORT

1-22. HR support includes the human resources functions of manning the force, HR services, personnel support, and HR planning and operations. HR support maximizes operational effectiveness and facilitates support to Soldiers, their Families, Department of Defense (DOD) civilians, and contractors who deploy with the force. HR support includes personnel readiness management (PRM); personnel accountability; strength reporting; personnel information management (PIM); casualty operations; EPS, band support, postal operations; reception, replacement, return-to-duty, rest and recuperation, and redeployment operations; morale, welfare, and recreation (MWR); and HR planning and staff operations. (see FM 1-0, FMI 1-0.01, and FMI 1-0.02).

Religious Support

1-23. Religious support facilitates the free exercise of religion, provides religious activities, and advises commands on matters of morals and morale. The First Amendment of the U.S. Constitution and Army Regulation (AR) 165-1 guarantee every American the right to the free exercise of religion. Commanders are responsible for those religious freedoms within their command. Chaplains perform and provide Religious Support (RS) in the Army to ensure the free exercise of religion (see FM 1-05).

Financial Management (FM) Operations

1-24. FM is comprised of two mutually supporting core functions: Finance and Resource Management operations. Finance operations include developing policy, providing guidance and financial advice to commanders; disbursing support to the procurement process; banking and currency; accounting; and limited pay support. RM operations include providing advice to commanders; maintaining accounting records; establishing a management internal control process; developing resource requirements; identifying, acquiring, distributing, and controlling funds; and tracking, analyzing, and reporting budget execution (see FM 1.06).

Legal Support

1-25. Legal support is the provision of professional legal services at all echelons. Legal support encompasses all legal services provided by judge advocates and other legal personnel in support of units, commanders, and Soldiers in an area of operation (AO) and throughout full spectrum operations. Judge Advocate General's Corps personnel assist Soldiers in personal legal matters and advise commanders on a wide variety of operational legal issues. These include the law of war, rules of engagement, lethal and nonlethal targeting, treatment of detainees and noncombatants, fiscal law, claims, contingency contracting, the conduct of investigations, and military justice. (see FM 1-04).

Band Support

1-26. Army bands provide critical support to the force by tailoring music support throughout military operations. Music instills in Soldiers the will to fight and win, foster the support of our citizens, and promote our national interests at home and abroad (see FM 1-0 and FM 1-19).

HEALTH SERVICES SUPPORT

1-27. Health services support is all support and services performed, provided, and arranged by the AMEDD to promote, improve, conserve, or restore the mental and physical well being of personnel in the Army and, as directed in other Services, agencies and organizations. Army Health System (AHS) support includes both HSS and force health protection (FHP). The HSS mission is a part of the sustainment WFF. The FHP mission falls under the protection WFF, but will be included to provide an accurate description of AHS support. This includes casualty care (encompassing a number of AMEDD functions—organic and area medical support, hospitalization, the treatment aspects of dental care and behavioral health (BH)/neuropsychiatric treatment, clinical laboratory services, and the treatment of chemical, biological, radiological, and nuclear [CBRN] patients), medical evacuation, and medical logistics. See FM 4-02.12 for a full description of AHS support.

OTHER SUSTAINMENT RELATED FUNCTIONS

1-28. As a result of the movement from battlefield operating systems to the WFF construct, some tasks are realigned. Two of those tasks are explosive ordnance disposal and Interment/Resettlement operations.

EXPLOSIVE ORDNANCE DISPOSAL

1-29. Explosive Ordnance Disposal is the detection, identification, on-site evaluation, rendering safe, recovery and disposal of explosive ordnance (EO) /improvised explosive devices (IEDs), weapons of mass destruction (WMD) which threaten forces, citizens, facilities, critical infrastructure, or operations. The Army EOD mission is to support national security strategy and national military strategy by reducing or eliminating EO/IED/WMD during operations. From a WFF perspective, EOD falls under the Protection WFF (see FM 4-30.50 for details).

INTERNMENT AND RESETTLEMENT OPERATIONS

1-30. Internment and Resettlement (I/R) operations are included under the Sustainment WFF (FM 3.0). While not a major sub-function of the sustainment WFF; I/R are supported by logistics, personnel services, and HSS. The Army is the DOD executive agent (EA) for all detainee operations. Within the Army, and through the CCDR, the Military Police (MP) are tasked with coordinating shelter, protection, accountability, and sustainment for detainees. The I/R function addresses MP roles when dealing with detainees, dislocated civilians, and US military prisoners. The MPs support the battlefield commander by relieving him/her of the problem of handling detainees with combat forces. The MPs perform the internment and resettlement functions of collecting, evacuating, and securing detainees (see FM 3-19.1 Military Police Operations and FM 3-19.40, Internment/Resettlement Operations). Sustainment in support of internment and resettlement will be discussed in Chapter 4.

SUMMARY

1-31. Sustainment builds and maintains combat power and provides strategic and operational reach and endurance. Sustainment supports full spectrum operations. It is one of six Army WFFs. The eight principles of sustainment must be considered throughout the operations process to achieve successful support of full spectrum operations. The principles are not a checklist but a guide for planners and leaders to incorporate the commander's intent for sustainment throughout the operations process. The sustainment WFF consists of three major sub-functions: logistics, personnel services, and Army health systems support Chapter 5 includes a more detailed discussion of the functional elements of sustainment.

Chapter 2

Roles and Responsibilities for Providing Sustainment in Unified Action

Unified action is the synchronization, coordination, and/or integration of the activities of governmental and nongovernmental entities with military operations to achieve unity of effort (FM 3-0). Unified action involves the application of all instruments of national power, including actions of OGA and multinational military and non military organizations. It requires joint integration. All Services must effectively operate together. Sustainment of unified action is pivotal to achieving campaign end state. The combination of diverse sustainment capabilities generate and sustain combat power that is more potent than the sum of its parts. The integration of sustainment capabilities maximizes efficiencies in delivering support to campaigns. This chapter will discuss the roles and functions of the Army, joint, interagency, intergovernmental, and multinational organization's contribution to the sustainment of unified actions.

JOINT INTERDEPENDENCE

2-1. Joint interdependence is the purposeful reliance by one Service's forces on another Service's capabilities to maximize the complementary and reinforcing effects of both. Army forces operate as part of an interdependent joint force. One area of joint interdependence is joint sustainment. Sustainment is inherently joint. All services use logistics, personnel services, and health services to support their forces. The mutual reliance on joint sustainment capabilities makes for a more effective utilization of sustainment resources. Combinations of joint capabilities defeat enemy forces by shattering their ability to operate as a coherent, effective force.

SUSTAINMENT OF JOINT FORCES

2-2. Sustainment of joint forces is the deliberate, mutual reliance by each Service component on the sustainment capabilities of two or more Service components (FM 3-0). Effective sustainment determines the depth to which the joint force can conduct decisive operations; allowing the JFC to seize, retain, and exploit the initiative (JP 3-0). It provides JFCs with flexibility to develop any required branches and sequels and to refocus joint force efforts as required. Sustainment is crucial to supporting operations. CCDRs and their staffs must consider a variety of sustainment factors including defining priorities for sustainment and common user logistics (CUL) functions and responsibilities.

2-3. CUL is materiel or service support shared with or provided by two or more Services, DOD agencies, or multinational partners to another Service, DOD agency, non-DOD agency, and/or multinational partner in an operation (JP 4-07). It can be restricted by type of supply and/or service and to specific unit(s) times, missions, and/or geographic areas. Service component commands, DOD Agencies (such as Defense Logistics Agency (DLA)), and Army commands (such as USAMC), provide CUL to other service components, multinational partners, and other organizations (such as NGOs).

2-4. The Army Service Component Command (ASCC) is responsible for providing support to Army forces and common-user logistics to other Services as directed by the CCDR and other authoritative instructions. The TSC is the logistic C2 element assigned to the ASCC and is the Army logistic headquarters (HQ) within a theater of operations. When directed, the TSC provides lead Service and executive agency support for designated CUL to OGA, MNFs, and NGOs. Additionally, the MEDCOM(DS) provides AHS support to other services when directed.

TITLE 10 RESPONSIBILITY

2-5. Each Service retains responsibility for the sustainment of forces it allocates to a joint force. The Secretary of the Army exercises this responsibility through the Chief of Staff the Army (CSA) and the ASCC assigned to each combatant command. The ASCC is responsible for the preparation and administrative support of Army forces assigned or attached to the combatant command.

2-6. The ASCC is responsible for all Army Title 10 functions within the CCDR's AOR. When an ASCC is in support of a GCC, it is designated as a Theater Army. The Theater Army commander exercises administrative control (ADCON) over all Army forces within the CCDR's AOR. He/She is responsible for preparing, training, equipping, administering, and sustaining Army forces assigned to combatant commands.

2-7. Title 10, United States Code (USC) specifies that individual services retain responsibility for sustainment, but the purposeful combination of complimentary service capabilities to create joint interdependent forces is often the most effective and efficient means by which to sustain a joint force. Additionally, common user support may be controlled and provided by other means. Options for executing sustainment of a joint force include any combination of the following:

- **Executive agent (EA).** The Secretary of Defense (SECDEF) or the Deputy Secretary of Defense, may designate the head of a DOD component (such as Chief of a Service, CCDR, or director of a Combat Support Agency) as an EA for specific responsibilities, functions, and authorities to provide defined levels of support for operational missions, administrative, or other designated activities that involve two or more DOD components. By definition, the designation as an EA makes that organization responsible for a joint capability within the boundaries of the EA designation.
- **Lead Service.** The CCDR may choose to assign specific CUL functions, to include both planning and execution to a Lead Service. These assignments can be for single or multiple common user functions and may also be based on phases and/or locations within the AOR. The CCDR may augment the Lead Service logistics organization with capabilities from another component's logistics organizations as appropriate. The Lead Service must plan issue procedures and sustainment funding for all items issued to other Services as well as a method for collecting items from other Services.
- **Subordinate logistics command.** The CCDR may assign joint logistics responsibilities to a subordinate Service component and establish a joint command for logistics (see JP 4-0). In order for the subordinate logistics command to be successful, the CCDR must augment it with the capabilities needed to integrate and control the delivery of theater support to meet the joint force requirements. This joint functional component must also be capable of integrating personnel, requirements, processes, and systems from all the joint force components.

2-8. The Theater Army's ADCON of Army forces entails providing administrative (legal, human resources, and finance) and logistics support to Army forces. When designated as an EA, the Army also enters into inter-Service, interagency, and intergovernmental agreements for certain responsibilities. These may include:

- General engineering support.
- Common-user land transportation.
- Disaster assistance.
- Force protection.
- Mortuary services.
- Detainee operations.
- Bulk fuel management.
- Postal operations.

2-9. The Theater Army commander's principal focus is on operational-level theater support involving force generation and force sustainment during campaigns and other joint operations. The Theater Army commander matches sustainment requirements for a campaign to the capabilities of the Army forces. In all joint operations, sustainment is a service responsibility unless otherwise directed by EA directives, CCDR lead service designations, or interservice support agreements (ISSAs).

2-10. The Theater Army, in concert with their associated geographic CCDRs, is responsible for identifying sustainment requirements, coordinating responsibilities, and establishing requisite Army C2 for sustainment. Furthermore, the Theater Army commander is responsible for properly executing all Army lead service or ISSA and related common support requirements within the theater.

2-11. Title 10 also includes two other CCDR's responsibilities and authorities that overlap the military departments' Title 10 functions. These are joint training and directive authority for logistics (DAFL). We will only discuss DAFL.

Directive Authority for Logistics (DAFL)

2-12. DAFL is the CCDR authority to issue directives to assigned forces. It includes peacetime measures to ensure the effective execution of approved OPLANs, effectiveness and economy of operation, prevention or elimination of unnecessary duplication of facilities, and overlapping of functions among the Service component commands. The President or SecDef may extend this authority to attached forces when transferring forces for a specific mission, and should specify this authority in the establishing directive or order.

2-13. CCDRs exercise DAFL and may delegate directive authority for a common support capability. The CCDR may delegate directive authority for as many common support capabilities to a subordinate JFC as required to accomplish the subordinate JFC's assigned mission. When the CCDR gives a service component CUL responsibility, he/she must specifically define the responsibilities. For some commodities or support services common to two or more Services, one Service may be given responsibility for management based on DOD designations or ISSAs. However, the CCDR must formally delineate this delegated directive authority by function and scope to the subordinate JFC or Service component commander.

2-14. The CCDR may elect to assign responsibility to establish a joint command for logistics to a subordinate Service component. The senior logistics HQ of the designated Service component will normally serve as the basis for this command. This may be an organization joint by mission (such as campaigns, major operations, and humanitarian missions), but not by design. When exercising this option, the CCDR retains DAFL and must specify the control and tasking authorities being bestowed upon the subordinate joint command for logistics, as well as the command relationships it will have with the Service components.

Other Combatant Command Authority

2-15. Besides logistics, the CCDR has the authority to direct certain other functional activities. These activities include personnel services and health services support.

Personnel Services

2-16. The CCDR's command authority allows him/her to direct and approve those personnel services necessary to carry out assigned missions. It also allows for the standardization of personnel policies within the command.

Health Services Support (HSS)

2-17. The CCDR requires medical capabilities that are scalable to the requirement, interoperable with other medical forces, and capable of rapid deployment into the joint operations area (JOA). Each Service has organic medical units with capabilities that are tailored for their traditional roles and missions and are normally capable of meeting other Services' requirements. The joint force surgeon provides recommendations to the JFC on the effective employment of all HSS capabilities. The AHS provides support to other Services when directed by the CCDR.

2-18. The theater surgeon recommends a theater evacuation policy through the CCDR and the Joint Chiefs of Staff (JCS) for approval by the SECDEF. The evacuation policy establishes the number of days an injured or ill Soldier may remain in the theater before returning to duty.

STRATEGIC LEVEL PROVIDERS

2-19. Strategic providers enable U.S. forces to maintain combat power and enable strategic reach and ensure endurance. Strategic providers include DOD agencies and commands. They link the national base to the theater. These agencies and commands provide sustainment to joint and Army forces.

INDUSTRIAL BASE

2-20. The industrial base consists of privately owned and government-owned industrial capability and capacity for manufacture, maintenance, modification, overhaul, and/or repair of items required by the U.S. and selected allies. It includes the production base and maintenance base. Active plants and production lines have some capability to surge. Repair parts manufacturers may be able to surge production for items that sustain deployed weapon systems. National policy requires the use of commercial materiel as much as possible.

DEFENSE LOGISTICS AGENCY (DLA)

2-21. DLA provides support for military departments and the GCC during peace and war. DLA is the focal point for the industrial base and is the EA for consumable supply items. DLA procures, stores, and distributes items to support the military Services and other customers. It also buys and distributes hardware and electronic items used in the maintenance and repair of military equipment. Excluded supply items are munitions, missiles, and military Service unique items.

2-22. DLA provides contract administration services to all DOD components and administers and supervises:

- The Federal Catalog System.
- The Defense Personal Property Reutilization Program.
- The DOD Industrial Plant Equipment Reserve.
- The Defense National Stockpile.

2-23. DLA provides reutilization and marketing services at the strategic through operational levels. Initially, salvage and excess materiel destined for the Defense Reutilization and Marketing Office is collected in theater areas. As the theater matures, DLA-directed activities may use HN or contractor support to assist in retrograding this materiel for inspection, classification, and disposal.

U.S. TRANSPORTATION COMMAND (USTRANSCOM)

2-24. USTRANSCOM is responsible for providing common-user and commercial air, land, and sea transportation (including patient movement), terminal management, and aerial refueling to support global deployment, employment, sustainment, and redeployment of U.S. forces. This ability makes possible projecting and maintaining national power where it is needed with the required speed, agility, high efficiency, and accuracy. The USTRANSCOM commander has the authority to procure commercial transportation services (such as Logistics Civil Augmentation Program (LOGCAP)), through its transportation component commands and to activate, with approval of the SECDEF, the Civil Reserve Air Fleet and Ready Reserve Fleet.

2-25. USTRANSCOM is DOD's Distribution Process Owner (DPO). As the DPO, USTRANSCOM is responsible for monitoring and managing the global distribution network. The DPO ensures the flow of force movement and sustainment for the supported GCC.

2-26. USTRANSCOM is composed of three component commands which remain under the combatant command of USTRANSCOM in contingency operations; Air Mobility Command (AMC), Military Sealift Command (MSC), and the Military Surface Deployment and Distribution Command (SDDC). These component commands provide inter-modal transportation to meet our national security objectives. While these commands normally remain under the C2 of USTRANSCOM in contingency operations, operational command (OPCON) or tactical control (TACON) could be delegated to a theater upon request of the GCC and approval of Commander, USTRANSCOM.

Defense Finance and Accounting Service (DFAS)

2-27. The DFAS is responsible for the delivery of responsive accounting and finance services. DFAS is an agency supporting the Office of the Under Secretary of Defense-Comptroller, the principal advisor to the SECDEF for budgetary and fiscal matters. It is DFAS's responsibility to coordinate and collaborate with all civilian defense agencies, military services, and the combatant commands that provide warfighting capabilities.

U.S. Army Human Resources Command (USAHRC)

2-28. The USAHRC integrates, manages, monitors, and coordinates HRS to develop and optimize Army human resources across the spectrum of conflict. The commander of USAHRC is the Army functional proponent for the military personnel management system and operates within the objectives set by the Army G-1. USAHRC major functions include the following:

- Execute the nine major functional categories of the Army personnel life cycle: force structure, acquisition, individual training and development, distribution, deployment, sustainment, professional development, compensation, and transition.
- Man the force and provide personnel support and human resources services to Soldiers, their Families, and organizations.
- Synchronize all military personnel activities to achieve efficient and cost effective execution of all human resources processes on an Army-wide basis to ensure current and future personnel requirements are defined.
- Interact with human resource organizations, including U.S. Army training centers, U.S. Army garrisons, divisions and corps, installations, and forward deployed bases to ensure policy, procedures, and service delivery systems support operational requirements at all levels.

U.S. Army Finance Command (USAFINCOM)

2-29. The USAFINCOM is an operating agency of the Assistant Secretary of the Army (Financial Management & Comptroller) (ASA(FM&C)). USAFINCOM provides advice and management information to the ASA(FM&C) and interacts between the Army staff, Army commands, units, and DFAS on matters concerning finance and accounting policy, systems, procedures, and reporting.

GENERATING FORCE

2-30. The generating force consists of those Army organizations whose primary mission is to generate and sustain the operational Army's capabilities for employment by JFCs. The generating force activities include support of readiness, ARFORGEN, and the routine performance of functions specified and implied in Title 10. As a consequence of its performance of functions specified and implied by law, the generating force also possesses operationally useful capabilities for employment by or in direct support of JFCs. Generating force capabilities include analyzing, understanding and adapting, and generating operational forces tailored to the specific context in which they will be employed.

2-31. The generating force's ability to develop and sustain potent land power capabilities is useful in developing partner security forces and governmental institutions, with its capability to develop, maintain, and manage infrastructure. Moreover, an increasingly pervasive information environment, combined with improved transportation capabilities, allow the effective application of capabilities from outside the area of operations.

2-32. The generating force is responsible for moving Army forces to and from ports of embarkation. They also provide capabilities to assist in the management and operation of ports of embarkation and debarkation and provide capabilities to GCC to conduct reception, staging, onward movement, and integration (RSO&I).

2-33. Generating force organizations provide a continuum of support that integrates the sustainment base with operating forces (see Figure 2-1). For example, USAMC is part of the generating force. It deploys certain organizations forward with an operational focus and role, such as the Army Field Support Brigade (AFSB), Logistic Support Element, Army Field Support Battalion, Brigade Logistics Support Team, and Contracting Support Brigade (CSB).

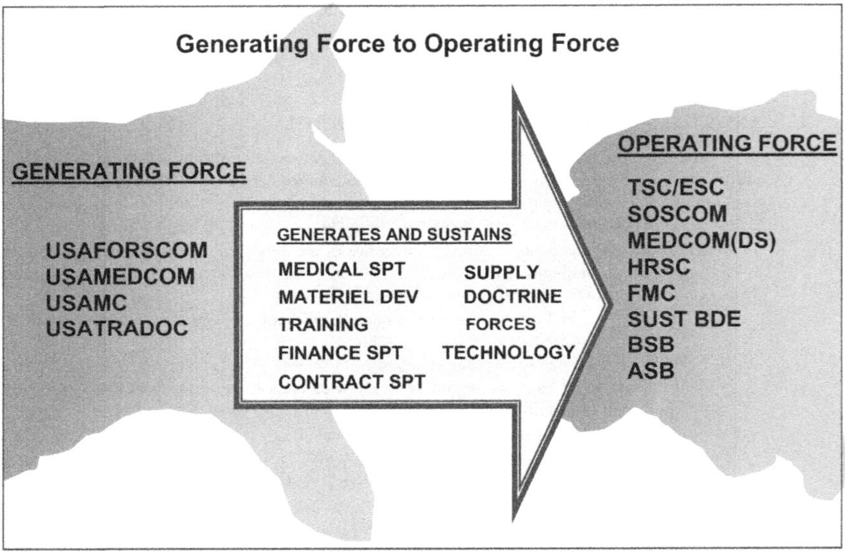

Figure 2-1. Generating Force Support to Operating Force

U.S. Army Training and Doctrine Command (USATRADOC)

2-34. USATRADOC is designated by the Secretary of the Army (SECARMY) as an Army command (ACOM) under the direction of Headquarters, Department of the Army (HQDA) and is assigned to carry out certain roles and functions of the SA pursuant to 10 USC 3013(b) with regard to Army forces worldwide. It recruits and trains uniformed personnel, manages the Army's education system, and operates the Army's training centers. TRADOC leads Army requirements determination; integrates doctrine, organization, training, materiel, leadership, and education' personnel and facilities (DOTLM-PF) developments to support required capabilities; and coordinates, synchronizes, and integrates Army capabilities developments with other ACOMs, the Combatant commands, the Joint Staff, and other military departments (AR 10-87).

2-35. TRADOC's training and education missions include preparing leaders of all branches and specialties (including sustainment branch leaders) for combined arms operations. Its doctrinal publications establish and support the Army operational concept. Sustainment leaders throughout the Army, but notably in the Sustainment Center of Excellence (SCOE), work with TRADOC to ensure the integration of sustainment related DOTLM-PF elements with Army operations. TRADOC's array of centers and schools develop comprehensive and integrated sustainment solutions for the Army through coordination with USAMC, U.S. Army Medical Department Center and School, Judge Advocate General's Legal Center and School, and the U.S. Army Chaplain Center and School.

U.S Army Materiel Command (USAMC)

2-36. USAMC equips and sustains the Army, providing strategic impact at operational speed. It provides logistics technology, acquisition support, contracting and contractor management, and selected logistics support to Army forces. It also provides related common support to other Services, multinational, and interagency partners. The capabilities of USAMC are diverse and are accomplished through its various major subordinate commands and other subordinate organizations.

2-37. USAMC is the lead for the Army's national-level maintenance and supply programs which are managed and executed by its subordinate Life Cycle Management Commands (LCMCs). These USAMC LCMCs coordinate with the USAMC staff as well as related ASA(ALT) Program Executive Officers (PEOs) and Product/Project Managers offices. Together, these USAMC LCMC and ASA(ALT) elements work to ensure support for fielded weapon systems and equipment for their entire life cycle. The LCMCs support to deploying and deployed forces is coordinated through the Army Sustainment Command (ASC) and is executed under the control of the supporting AFSB. LCMCs are discussed in more detail below.

2-38. In addition to the functions performed by the LCMCs, USAMC exercises overall responsibility of sustainment maintenance for the Army and managing secondary items through the National Maintenance Program, whose tenets are as follows:

- Managing sustainment maintenance unit workloads to meet national requirements.
- Ensuring all component repairs are performed to a national standard.
- Ensuring sustainment maintenance providers possess the facilities, tools, test, measurement, and diagnostic equipment, skills, and workforce required to meet national standards.
- Facilitating quality assurance by ensuring that maintainers use documented quality systems and are technically certified to repair to standards.

2-39. USAMC is also the lead, but not sole, Army organization responsible for providing contracting services to the Army. USAMC contracting support includes the LOGCAP. Through its subordinate contracting commands, USAMC provides both institutional and operational contract support planning assistance and contract execution support to all Army forces.

2-40. USAMC, through its subordinate organizations, determines the optimum sustainment maintenance capability to maximize supply flow within CONUS through the coordination between the TSC and AFSB in theater.

Materiel Enterprise (ME)

2-41. The ME executes Army-wide materiel lifecycle management. The Commanding General, USAMC chairs the Enterprise Governance Board with oversight provided by the ASA (ALT). The ME develops and then uses the approved Army's Equipping Strategy to deliver and sustain equipment readiness. It also advises the SECARMY on Army-wide materiel issues to enhance materiel readiness. The ME supports ARFORGEN by providing the most effective and sustainable equipment for Soldiers.

USAMC Director of Materiel (DOM)

2-42. The Army's Enterprise Initiative aligns ammunition, supply, and maintenance functions formerly performed by the Installation Management Command (IMCOM), Director of Logistics (DOL) to USAMC as the Director of Materiel (DOM). This alignment assures the efficiencies and core competencies of the ME. USAMC DOMs workload the ammunition, supply, and maintenance functions as an integral part of the industrial base in order to sustain equipment readiness in support of the ARFORGEN process.

USAMC Subordinate Commands

2-43. USAMC has several subordinate commands. These commands are discussed below.

Surface Deployment and Distribution Command (SDDC)

2-44. SDDC is a subordinate command of USAMC and the ASCC of U.S. TRANSCOM. It provides inter-modal transportation to meet national security objectives.

The Aviation and Missile Life Cycle Management Command (AM LCMC)

2-45. The AM LCMC integrates functions across their commodity and sustains aviation, missile, and unmanned aerial vehicle systems, ensuring weapons systems readiness with seamless transition to combat operations. It assists materiel developers (PEO/PM) with the development, acquisition, and fielding of aviation, missile systems, and related equipment. The LCMC performs applied research, integrated logistics support, materiel readiness management, and maintenance support for Army aviation, missile weapon systems, subsystems, and associated equipment.

The CECOM Life Cycle Management Command (CECOM LCMC)

2-46. The CECOM LCMC integrates functions across their commodity and sustains command, control, communications, computers, intelligence, surveillance, and reconnaissance (C4ISR) information systems for joint interoperability. It assists materiel developers (PEO/PM) with the development, acquisition, and fielding of C4ISR systems. The LCMC performs applied research, integrated logistics support, materiel readiness management, and maintenance support and provides technical support capabilities to deploying and deployed Army forces.

The Tank-Automotive and Armaments Life Cycle Management Command (TA LCMC)

2-47. The TA LCMC integrates functions across their commodity and sustains Soldier and ground systems for the operating force. This is accomplished through the integration of effective and timely acquisition, logistics, and technology. The TA LCMC is responsible for integration of initial fielding requirements with current item sustainment. It provides acquisition support of TACOM commodity end items, spare parts, and supplies for U.S. and Allied weapon systems and overhauls, modernizes, and repairs TACOM commodity equipment.

The Joint Munitions and Lethality Life Cycle Management Command (JM&L LCMC)

2-48. The JM&L LCMC manages research, development, production, storage, distribution, and demilitarization of conventional ammunition. The JM&L LCMC consists of the Program Executive Office for Ammunition, the Joint Munitions Command (JMC), and the Armament, Research, Development and Engineering Center (ARDEC). The JM&L LCMC serves as the Single Manager for Conventional Ammunition (SMCA) and serves as the SMCA Field Operating Activity. PEO Ammunition develops and procures conventional and leap ahead munitions to increase combat firepower to execute total ammunition requirements. ARDEC is the principal researcher, technology developer, and sustainer of current and future munitions. JMC serves as the logistics and readiness arm; producing, storing, issuing, and demilitarizing conventional ammunition for all U.S. Services, other U.S. agencies, and allied nations. The JMC also provides munitions training through its Defense Ammunition Center for Quality Assurance Specialist (Ammunition Surveillance) and Ammunition Manager Career programs as well as explosive safety training for all munitions handlers.

U.S. Army Sustainment Command (ASC)

2-49. The ASC is responsible for coordinating generating force support to the operating force. The ASC works in close coordination with other USAMC and national level sustainment and distribution organizations such as DLA and USTRANSCOM and the respective deployed TSCs. The ASC executes its operating force mission through its deployable AFSBs. The AFSB integrates system support contracting into the overall theater support plan. Theater support contracting, to include LOGCAP, is planned, coordinated, and executed by the ASCC supporting CSB under the C2 of U.S. Army Contracting Command's (USACC) Expeditionary Contracting Command (ECC). The ASC provides continuous equipment and materiel readiness to CONUS forces through planning, resourcing, and materiel management in accordance with the ARFORGEN process. This is achieved by synchronizing strategic materiel management and by integrating acquisition, logistics, and technology.

2-50. In conjunction with LCMCs, the ASC through its DMC, workloads maintenance, supply, and ammunition functions at U.S. Army installations. It may also workload Field Logistics Readiness Centers, depots, and contractors to support ARFORGEN. The ASC also integrates the Logistics Assistance Program (LAP) in support of contingency operations. Most importantly, the ASC generates and projects combat power to support expeditionary operations.

U.S. Army Contracting Command (USACC)

2-51. The SECARMY directed the establishment of the USACC to consolidate most Army contracting under a single command. This includes most of the active component contingency contracting force structure. This change in structure represents a fundamental change in the C2 and support relationships of contingency contracting force structure. USACC provides contracting support through its two subordinate commands: the ECC and the Mission and Installation Contracting Command (MICC). In addition, the USACC provides reach back contracting support from its CONUS based acquisition centers. The ECC is responsible for theater support contracting in support of deployed forces and garrison contracting in support of all OCONUS Army installations and associated forward station units. The ECC also coordinates LOGCAP planning and assists in the integration of LOGCAP support through its CSBs. The MICC, through its continental Unites States (CONUS) director of contracting offices, is responsible for providing supplies, services, and construction contracts in support of all CONUS installations.

2-52. The CSB is a Table of Organization and Equipment (TOE) unit of USAMC assigned to the USACC. The CSB consolidates all theater support contracting capabilities into one command. The CSB serves under the C2 and procurement authority of the ECC, providing direct support to the theater or mission TSC as described in their contracting support plan when deployed. The CSB commander is also the ASCC Principal Assistant Responsible for Contracting (PARC). Like the AFSB, the CSB is regionally aligned and provides theater support contracting and planning in support of the ASCC, Army forces, and their subordinate commands.

U.S. Army Research Development and Engineering Command (RDECOM)

2-53. The RDECOM rapidly transitions state of the art technology to the force. RDECOM develops supplies and equipment from combat rations, clothing, battledress, to weapons, vehicles, and future combat systems for the force.

U.S. Army Security Assistance Command (USASAC)

2-54. USASAC manages Army security assistance that provides program management, including planning, delivery, and life cycle support of equipment, services, and training to and co-production with U.S. allies and multinational partners.

U.S. Army Chemical Materials Agency (USACMA)

2-55. The USACMA provides safe and secure storage of the chemical stockpiles and recovered chemical warfare material; and destroys all chemical warfare materials.

U.S. Army Medical Command (USAMEDCOM)

2-56. The USAMEDCOM provides Army health support for–mobilization, deployment, sustainment, redeployment, and demobilization. USAMEDCOM is a direct reporting unit. The Surgeon General is the Commander, USAMEDCOM. The USAMEDCOM provides enterprise-level AHS support to the joint force, ensuring complete continuity of care. The USAMEDCOM integrates the capabilities of its subordinate operational Army medical units with generating force assets such as medical treatment facilities (MTFs) and research, development, and acquisition capabilities. The USAMEDCOM's generating force capabilities not only augment those of operating forces but also provide significant assistance in coping with unanticipated health threats.

2-57. The USAMEDCOM simultaneously maintains the capability to provide continuity of patient care, while ensuring it retains the ability to care for patients returning from theaters. The USAMEDCOM also provides individual AMEDD training and medical materiel and research and development activities to support the Army mobilization force. The USAMEDCOM's strategic AHS Support focuses on:

- US Army Dental Command.
- US Army Medical Research and Materiel Command.
- US Army Veterinary Command.
- US Army Medical Department Center and School.
- US Army Center for Health Promotion and Preventive Medicine.

2-58. The USAMEDCOM also has regional medical commands responsible for oversight of day-to-day operations in military treatment facilities, exercising C2 over the MTFs in the supported region. See FM 4-02 and FM 1-01 for additional information.

U.S. Army Forces Command (USAFORSCOM)

2-59. The USAFORSCOM is the Army component of U.S. Joint Forces Command. Joint Forces Command's mission is to provide U.S. military forces where needed throughout the world and to ensure they are integrated and trained as unified forces ready for any assigned task. The FORSCOM commander functions as commander of the Army forces of this unified command and provides military support to civil authorities, including response to natural disasters and civil emergencies.

2-60. FORSCOM trains, mobilizes, deploys, and sustains combat ready forces capable of responding rapidly to crises worldwide. FORSCOM develops and cares for Soldiers, optimizes resources, develops installations, and establishes facilities to project power globally.

OPERATING FORCE

2-61. The operating forces are those forces whose primary missions are to participate in combat and the integral supporting elements thereof (see FM1-01). By law, operational Army units are typically assigned to CCDRs. The Army normally executes its responsibilities to organize, train, and equip operational Army units through ASCCs.

Army Service Component Command (ASCC)/Theater Army (TA)

2-62. When an ASCC is in support of a GCC, it is designated as a Theater Army. The TA is the primary vehicle for Army support to joint, interagency, intergovernmental, and MNFs. The TA HQ performs functions that include RSO&I; logistics over-the-shore operations; and security coordination.

2-63. The TA exercises ADCON over all Army forces in the area of responsibility unless modified by DA. This includes Army forces assigned, attached, or OPCON to the combatant command. As such, the TSC is assigned to the TA. The TA coordinates with the TSC for operational sustainment planning and management. The TA defines theater policies and coordinates with the TSC for technical guidance and execution of force projection and sustainment.

Corps

2-64. The corps provides a HQ that specializes in operations as a land component command HQ and a joint task force for contingencies. When required, a corps may become an intermediate tactical HQ under the land component command, with OPCON of multiple divisions (including multinational or Marine Corps formations) or other large tactical formations.

2-65. The corps HQ is designed to, in priority, C2 Army forces, leverage joint capabilities, and C2 joint forces for small-scale contingencies. Its primary mission is to C2 land forces in land combat operations.

The corps HQ has the capability to provide the nucleus of a joint HQ. However, the ability of the corps to transition to a joint task force (JTF) HQ or joint force land component command (JFLCC) HQ is heavily dependent on other Service augmentation. The transition of a modular corps HQ to a joint HQ relies on a timely fill of joint positions, receipt of joint enabling capabilities, and comprehensive pre-activation training as a joint HQ.

DIVISION

2-66. Divisions are the Army's primary tactical warfighting HQ. Their principal task is directing subordinate brigade operations. Divisions are not fixed formations. Therefore, they may not have all types of BCTs in an operation or they may control more than one of a particular type of BCT. A division can control up to six BCTs with additional appropriate supporting brigades during major combat operations. The types of support brigades are combat aviation, fires, maneuver enhancement, battlefield surveillance, and sustainment. The Sust Bde normally remains attached to the TSC but supports the division. The division may have OPCON of a Sust Bde while conducting large-scale exploitation and pursuit operations.

BRIGADE COMBAT TEAM (BCT)

2-67. As combined arms organizations, BCTs form the basic building block of the Army's tactical formations. They are the principal means of executing engagements. Three standardized BCT designs exist: heavy, infantry, and Stryker. Battalion-sized maneuver, fires, reconnaissance, and Brigade Support Battalion (BSB) are organic to BCTs.

THEATER SUSTAINMENT COMMAND (TSC)

2-68. **The TSC serves as the senior Army sustainment HQ for the Theater Army.** The TSC provides C2 of units assigned, attached, or under its OPCON. The mission of the TSC is to provide theater sustainment (less medical). See FM 4-93.4 for more information on the TSC.

2-69. The TSC is capable of planning, preparing, executing, and assessing logistics and human resource support for Army forces in theater or JFC. It provides support to full spectrum operations. As the distribution coordinator in theater, the TSC leverages strategic partnerships and joint capabilities to establish an integrated theater-level distribution system that is responsive to Theater Army requirements. It employs Sust Bdes to execute theater opening (TO), theater sustainment, and theater distribution operations.

2-70. The TSC includes units capable of providing multifunctional logistics: supply, maintenance, transportation, petroleum, port, and terminal operations. Other specialized capabilities, such as MA, aerial delivery, human resources, sustainment to I/R operations, and FM, are available from the force pool. The combination of these capabilities gives the TSC commander the ability to organize and provide tailored support.

EXPEDITIONARY SUSTAINMENT COMMAND (ESC)

2-71. Expeditionary Sustainment Commands (ESC) are force pooled assets that, under a command and control relationship with the TSC, command and control sustainment operations in designated areas of a theater. The ESC plans, prepares, executes, and assesses sustainment, distribution, theater opening, and reception, staging, and onward movement operations for Army forces in theater. It provides operational reach and span of control. It may serve as a basis for an expeditionary joint command when directed by the GCC or designated multinational or joint task force commander. It normally deploys to provide command and control when multiple Sust Bdes are employed or when the TSC determines that a forward command presence is required. This capability provides the TSC commander with the regional focus necessary to provide effective operational-level support to Army or JTF missions.

Human Resource Sustainment Center (HRSC)

2-72. **The HRSC is a multifunctional, modular organization (staff element) assigned to a TSC that provides HR support to the theater.** The HRSC integrates and ensures execution of HR support for postal, casualty, and personnel accountability operations throughout the theater as defined by the policies and priorities established by the ASCC G-1. The HRSC has a defined role to ensure that the theater HR support plan is developed and then supported with available resources.

2-73. The HRSC provides operational planning and technical support to the DMC in the TSC and support operations (SPO) in the ESC. The HRSC also provides technical guidance to the HR Operations Branch in Sust Bdes and ESCs, HR companies and teams, and theater S-1s/G-1s (see FMI 1-0.02).

Financial Management Center (FMC)

2-74. The FMC Director, in coordination with the Theater Army G8, is the principal advisor to the commander on all aspects of finance operations. The FMC is assigned to the TSC and provides technical oversight and coordination of all theater finance operations to include companies and detachments in theater. The FMC negotiates with HN banking facilities, is the U.S. Treasury account holder for central funding, advises unit commanders on the use of local currency, and coordinates with national providers (U.S. Treasury, DFAS, ASA(FM&C), and USAFINCOM) to establish FM support requirements. The FMC sustains Army, joint, and combined operations by providing technical oversight to timely contractual and procurement payments and theater disbursing capability (see FM 1-06).

Army Field Support Brigade (AFSB)

2-75. AFSBs are assigned to the ASC. **An AFSB provide integrated and synchronized acquisition logistics and technology (ALT) support, less medical, to Army operational forces.** AFSBs are regionally aligned to a Theater Army and will normally be in direct support to the TSC or the lead theater logistics commander. AFSBs serve as ASC's link between the generating force and the operational force. They are responsible for the integration of ALT capabilities in support of operational and tactical commanders across full spectrum operations. The AFSB integrates theater support contracting into the overall ALT support plan, in coordination with the USA ECC CSB Commander/Principal Assistant Responsible for Contracting (PARC), supporting the ASCC. This includes coordinating for ALT strategic reach capabilities via a technical reach or call-forward process (see FM 4-93.41).

Sustainment brigade (Sust Bde)

2-76. When deployed, the Sust Bde is a subordinate command of the TSC, or by extension the ESC. **The Sust Bde is a flexible, multifunctional sustainment organization, tailored and task organized according to METT-TC.** It plans, prepares, executes, and assesses sustainment operations within an area of operations. It provides C2 and staff supervision of sustainment operations and distribution management. Selected functions previously performed by Army of Excellence corps and division support commands and area support groups are consolidated with the Sust Bde. As a result it serves as a single operational echelon providing operational level C2 of logistics support (see FM 4-93.2).

2-77. Sust Bdes are primarily employed in a support relationship. Under certain METT-TC conditions, they may be OPCON to the Army forces commander when operating as the senior sustainment command or TACON for operational area security or other types of operations.

Combat Sustainment Support Battalion (CSSB)

2-78. **The CSSB is a flexible and responsive unit that executes logistics throughout the depth of an area of operations including transportation, maintenance, ammunition, supply, MA, airdrop, field services, water, and petroleum.** The CSSB is attached to a Sust Bde and is the building block upon which the Sust Bde capabilities are developed. The CSSB is tailored to meet specific mission requirements. Employed on an area basis, the CSSB plans, prepares, executes, and assesses logistics operations within an area of operations. The CSSB also supports units in or passing through its designated area.

2-79. The CSSB may operate remotely from the Sust Bde and therefore must maintain communications with the Sust Bde. The CSSB establishes voice communications to support C2 and convoy operations as well as to monitor, update, and evaluate the logistics posture (see FM 4-93.2).

Medical Command (Deployment Support) (MEDCOM [DS])

2-80. The MEDCOM (DS) serves as the senior medical command within the theater in support of the CCDR. The MEDCOM (DS) provides the medical C2 for medical units delivering health care in support of deployed forces. The MEDCOM (DS) is a regionally focused command and provides subordinate medical organizations to operate under the MEDBDE and/or MMB and forward surgical teams or other augmentation required by supported units.

2-81. The MEDCOM (DS) is a versatile, modular medical C2 structure composed of a main command post (MCP) and an operational command post (OCP). The operational command post provides medical C2, policy development, and technical guidance for subordinate medical units. It also provides interface and liaison with supported theater forces at echelons above brigade and subordinate BCTs. The MCP is a deployable augmentation module which completes the staffing of the MEDCOM (DS) HQ. The MCP facilitates medical C2, policy development, and technical guidance to subordinate MEDBDEs and provides interface and liaison with supported theater forces. See FM 4-02.12 for more information.

Medical Brigade (MEDBDE)

2-82. The MEDBDE provides a scalable expeditionary medical C2 capability for assigned and attached medical functional organizations task-organized for support of the BCTs and supported units at echelons above brigade (EAB). The MEDBDE provides all of the medical C2 and planning capabilities necessary to deliver responsive and effective AHS support. The MEDBDE ensures the right mixture of medical professional (operational, technical, and clinical) expertise to synchronize the complex system of medical functions.

2-83. The MEDBDE has the capability to provide an early entry module, an expansion module, and the campaign module, thus enabling its capability to be tailored to METT-TC factors of a specific operation. As the supported forces grow in size and complexity, the MEDBDE can deploy additional modules that build upon one another to support full spectrum operations. When required, a MMB may be employed to provide medical C2 and operational planning for task-organized medical functional teams, detachments, and companies.

Multifunctional Medical Battalion (MMB)

2-84. The MMB is designed as a multifunctional HQ. It provides medical C2, administrative assistance, medical logistics (MEDLOG) support, and technical supervision for assigned and attached medical functional organizations (companies, detachments, and teams) task-organized for support of a division and its BCTs. It can also be deployed to provide medical C2 to expeditionary forces in early entry operations and facilitate the RSO&I of theater medical forces. All EAB medical companies, detachments, and teams in theater may be assigned, attached, or placed under the OPCON of an MMB. The MMB is under the C2 of the MEDBDE/MEDCOM (DS).

Sustainment Brigade (Special Operations) Sust Bde (SO)

2-85. The Sust Bde (SO) is a subordinate command of the U.S. Army Special Operations Command. Its mission is to provide limited sustainment, FHP, and signal support to Army Special Operations Forces (ARSOF). ARSOF are not logistically self-sufficient. ARSOF units rely upon the GCC theater infrastructure for virtually all of their support above their organic capabilities. The planning and execution of logistics support to ARSOF must be nested within the GCC's concepts of operation and support, as well as tailored to interface with the theater logistics structures. For further information on ARSOF logistics capabilities refer to FM 3-05.140.

BRIGADE SUPPORT BATTALION (BSB)

2-86. BSBs are organic components of BCT, Fires, and Maneuver Enhancement Brigades. The BSB is tailored to support the particular brigade to which it is organic. For example, the BSB of a heavy brigade combat team (HBCT) has more fuel distribution capabilities and maintenance than does a fires brigade BSB. The battalion provides supply, maintenance, motor transport, and medical support to the supported brigade. The BSB plans, prepares, and executes, logistics operations in support of brigade operations.

AVIATION SUPPORT BATTALION (ASB)

2-87. The ASB is the primary aviation logistics organization organic to combat aviation brigade (CAB) and the theater aviation brigade. The ASB performs the BSB mission. Combat aviation brigades typically conduct attack, reconnaissance, security, movement to contact, air assault, air movement, aero medical evacuation, personnel recovery, and C2 support missions. It provides aviation and ground field maintenance, brigade-wide satellite signal support, replenishment of all supplies, and medical support to the aviation brigade. The ASB has been optimized to support the CAB's forward support companies, aviation maintenance companies, and the brigade HQ and HQ company (see FM 4-90.23).

INTERAGENCY COORDINATION

2-88. Interagency coordination is the coordination that occurs between elements of Department of Defense and engaged U.S. Government agencies for the purpose of achieving an objective (FM 3-0). It is an essential characteristic of unified action. Military operations must be coordinated with the activities of other agencies of the United States government, IGO, NGO, and activities of various HN agencies.

2-89. The SECDEF may determine that it is in the national interest to task U.S. military forces with missions that bring them into close contact with (if not in support of) IGOs and NGOs. In such circumstances, it is mutually beneficial to closely coordinate the activities of all participants. Unity of effort between IGOs, NGOs, and military forces should be the goal. Taskings to support IGOs and NGOs are normally for a short-term purpose due to extraordinary events. In most situations, sustainment, communications, mobility, and security are the capabilities most needed.

2-90. For some operations, sustainment forces may be employed in quantities disproportionate to their normal military roles and in nonstandard tasks. Furthermore, they may precede other military forces or may be the only forces deployed. They also may have continuing responsibility after the departure of combat forces in support of MNFs, OGAs, IGOs, or NGOs. In such cases, they must adhere to any applicable status-of-forces agreements (SOFA) and acquisition cross service agreements (ACSAs) to which the United States is a party.

2-91. In a national emergency or complex contingency operation, DOD and the U.S. military often serve in a supporting role to other agencies and organizations. Commanders and their staffs should develop an understanding of how military operations and capabilities can be coordinated with those of other agencies and organizations to focus and optimize the military's contributions to accomplish the desired end state. US agencies, the UN, IGOs, NGOs, and MNFs provide for their own sustainment. However, US military sustainment capabilities are frequently requested and provided to these organizations. The JTF may be asked to assume all or part of the burden of logistics for these organizations after arrival. This support may include intertheater and intratheater airlift, ground transportation of personnel, equipment, and supplies, airfield control groups, and port and railhead operations groups.

MULTINATIONAL SUSTAINMENT OPERATIONS

2-92. A major objective when Army forces participate in the sustainment of multinational deployments is to maximize operational effectiveness. Support provided and received in multinational operations must be in accordance with existing legal authorities. There are two types of multinational operations; alliances and coalitions.

ALLIANCE

2-93. An alliance is the relationship that results from a formal agreement (such as a treaty) between two or more nations for broad, long-term objectives that further the common interests of the members (see JP1-02). The North Atlantic Treaty Organization (NATO) is an example of an alliance. An alliance may use an integrated staff, instead of merely augmenting the staff of one nation's organization with other national representatives. Each primary staff officer could be a different nationality and usually the deputy commander represents a major participant other than the lead nation.

COALITION

2-94. A coalition is an ad hoc arrangement between two or more nations for common action (see JP1-02). Many coalitions are formed under the guidance of the United Nations (UN). The UN does not have a military organization; and therefore, no preplanned formal military structures. The American, British, Canadian, Australian, and New Zealand (ABCA) Armies Program represent a coalition of English speaking nations. ABCA forces have never been employed under the program however. However, the ABCA nations have served together in ad hoc coalitions on numerous occasions to pursue common objectives (ABCA Publication 332). Figure 2-2 demonstrates the differences between an alliance and coalition.

2-95. In multinational operations, sustainment of forces is primarily a national responsibility. However, relations between the United States and its NATO allies have evolved to where sustainment is viewed as a collective responsibility (NATO Military Committee Decision [MCD] 319/1). In multinational operations, the multinational commander must have sufficient authority and control mechanisms over assets, resources, and forces to effectively achieve his/her mission. For each nation of an alliance or coalition to perform sustainment functions separately would be inefficient and expensive. It would also hinder the multinational commander's ability to influence and prioritize limited resources to support his/her operation and accomplish his/her mission.

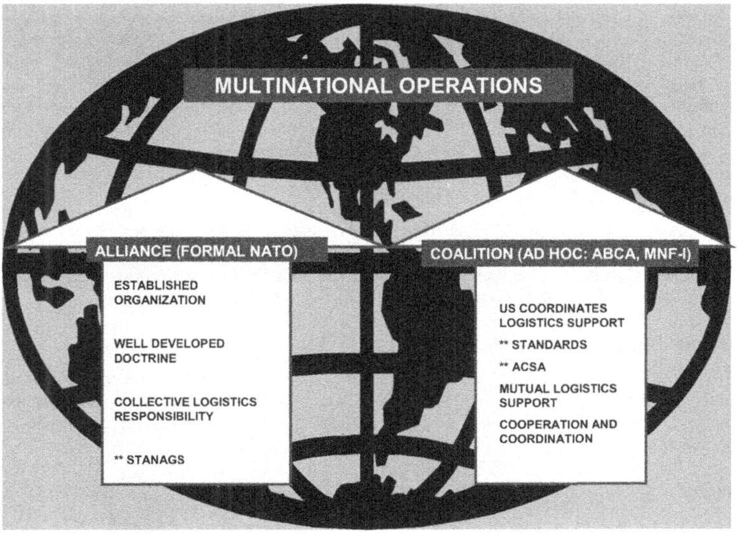

Figure 2-2. Multinational Operations

NATO Logistics Options

2-96. NATO doctrine allows for the formation of a Combined Joint Force Land Component Command (CJFLCC). The CJFLCC HQ can be set at a sub-regional command level or formation level. The CJFLCC commander establishes requirements and sets priorities for support of forces in accordance with the overall direction given by the Joint Force Commander. The commander coordinates sustainment operations with all participating nations.

2-97. Merging national sustainment systems into multinational support systems requires the willingness to share the control of vital support functions with a NATO commander and requires technical interoperability of national support assets. STANAGs provide agreed policy and standards to NATO nations and contribute to the essential framework for specific support concepts, doctrine procedures, and technical designs. Non-NATO nations will be expected to comply with NATO publications while on NATO-led operations.

2-98. The basic sustainment support options for multinational operations may range from totally integrated multinational sustainment forces to purely national support. NATO Allied Publication 4.2 provides details on the following support options.

National Support Element (NSE)

2-99. A NSE is any national organization or activity that supports national forces that are a part of a MNF. Their mission is nation-specific support to units and common support that is retained by the nation. It should also be noted that NSEs operating in the NATO commander's AO are subject to the SOFA, memorandums of agreements, and other HN arrangements.

Host Nation Support (HNS)

2-100. HNS is civil and military assistance rendered by a nation to foreign forces within its territory during peacetime, crises or emergencies, or war based on agreements mutually concluded between nations. Many HNS agreements have already been negotiated between NATO nations. Potential HNS agreements may address labor support arrangements for port and terminal operations, using available transportation assets in country, using bulk petroleum distribution and storage facilities, possible supply of Class III (Bulk) and Class IV items, and developing and using field services. The U.S. initiates and continually evaluates agreements with multinational partners for improvement. They should be specifically worded to enable planners to adjust for specified requirements. Additionally, the commander should assess the risk associated with using HNS, considering operational area security and operational requirements.

Contracting Support to Multinational Operations

2-101. A deployed force may be required to set up contractual arrangements with local (and non-local) contractors. These are normally negotiated individually with vendors to make use of whatever resources are available. Coordination between contributing nations and the in-place NATO HQ in their contractual dealings is essential. Coordination should be accomplished at the highest appropriate level.

Multinational Integrated Logistics Units (MILU)

2-102. A MILU is formed when two or more nations agree, under OPCON of a NATO commander, to provide logistics support to a MNF. MILUs are designed to provide specific logistics support where national forces cannot be provided, or could be better utilized to support the commander's overall sustainment plan.

Lead Nation

2-103. A Lead Nation for Logistic Support has agreed to assume overall responsibility for coordinating and/or providing an agreed spectrum of sustainment for all or part of a MNF within a defined geographical area. This responsibility may also include procurement of goods and services with compensation and/or reimbursement subject to agreements between the parties involved.

Role Specialisation

2-104. One nation may assume the responsibility for providing or procuring a particular class of supply or service for all or part of the MNF. A Role Specialist Nation's responsibilities include the provision of assets needed to deliver the supply or service. Compensation and/or reimbursement will then be subject to agreement between the parties involved.

Third Party Logistic Support Services (TPLSS)

2-105. TPLSS is the use of preplanned civilian contracting to perform selected sustainment. Its aim is to enable competent commercial partners to provide a proportion of deployed sustainment so that such support is assured for the commander and optimizes the most efficient and effective use of resources. TPLSS is most likely to be of use in a non-Article 5 Crisis Response Operation, and especially once the operational environment has become more benign. The TPLSS database, which NATO Maintenance & Supply Agency (NAMSA) developed, contains details of potential contractors world wide, capable of providing sustainment to NATO operations. NATO Commands and Nations may consider using the technical expertise of NAMSA for their contract activities.

Mutual Support Agreements (MSA)

2-106. Participating nations have the option to develop mutual support arrangements (bi- and multi-laterally) to ensure provision of logistic support to their forces. This is especially useful when nations have small force contingents collocated with the forces of another nation that have the capacity to support them. By working together and sharing resources (especially services capabilities), nations can achieve economies of scale in their sustainment operations. MSAs have the advantage of being simple to set up and can take place on an ad hoc basis.

OTHER SUSTAINMENT OPTIONS

2-107. Chapter 138 of Title 10 USC authorizes exchanging support between U.S. services and those of other countries. It authorizes DOD acquisition from other countries by payment or replacement-in-kind, without establishing a cross-servicing agreement. Supplies and services authorized under Chapter 138 do not include major end items, missiles, or bombs. It does include food, billeting, petroleum, oils, transportation, communication services, medical services, ammunition, storage, spare parts, maintenance services, and training. Therefore, negotiations in advance of operations for sharing projection and sustainment resources are recommended.

ACQUISITION CROSS-SERVICING AGREEMENT (ACSA)

2-108. Under ACSA authority (Title 10 USC, sections 2341 and 2342), the SECDEF can enter into agreements for the acquisition or cross-service of logistics support, supplies, and services on a reimbursable, replacement-in-kind, or exchange-for-equal-value basis. These agreements can be with eligible nations and international organizations of which the United States is a member. An ACSA is a broad overall agreement, which is generally supplemented with an implementing agreement (IA). The IA contains points of contact and specific details of the transaction and payment procedures for orders for logistics support. Neither party is obligated until the order is accepted.

2-109. Under these agreements, common logistics support includes food, billeting, transportation (including airlift), petroleum, oils, lubricants, clothing, communications services, medical services, ammunition, base operations, storage services, use of facilities, training services, spare parts and components, repair and maintenance services, calibration services, and port services. Items that may not be acquired or transferred under the ACSA authority include weapon systems, major end items of equipment, guided missiles, nuclear ammunition, and chemical ammunition (excluding riot control agents).

SUMMARY

2-110. Sustainment is joint interdependent. The deliberate and mutual reliance on joint sustainment can reduce duplication and increase efficiency. U.S. Title 10 requires each Service to provide its own logistics support. However, authority is available through other means to conduct joint sustainment. DAFL gives the CCDR authority to issue directives to subordinate commanders in order to meet joint sustainment needs. Strategic level sustainment is provided by joint organizations such as DLA, USTRANSCOM, USAHRC, DFAS, and USAFINCOM. These generating forces unique capabilities allow them to provide support directly to the operational Army's modular forces. Multinational operations may include coalitions and alliances. An alliance is a formal relationship between multinational partners whereas the coalition is an ad hoc relationship. Under the NATO alliance several options are available for providing sustainment to allied and coalition forces.

Chapter 3

Command and Control (C2)

C2 is the exercise of authority and direction by a properly designated commander over assigned and attached forces in the accomplishment of a mission. C2 is an art and a science. Commanders combine the art of command and the science of control to accomplish missions. This chapter will discuss C2, mission command, and the art of battle command from a sustainment perspective. Although the processes of C2 are the same for all commands, the mission focus, knowledge, understanding, and visualization of how support will be provided is different. C2 is fundamental to any discussion of sustainment because of the importance and flexibility of the modular force. The linking of C2 systems enables effective decision making as people, materiel, and medical support moves from generating force to operating force units. It also affects how commanders visualize, describe, and direct support. Because of the uncertain and ever changing nature of operations, mission command—as opposed to detailed command—is the preferred method for exercising C2. The use of mission orders, full familiarity with the commander's intent and concept of operations, and mutual trust and understanding between commanders and subordinates are prerequisites for mission command.

COMMAND

3-1. Command is the authority that an armed forces commander lawfully exercises over subordinates by virtue of rank or assignment. Command includes the authority and responsibility for effectively using available resources and for planning the employment of, organizing, directing, coordinating, and controlling military forces for accomplishment of assigned missions (see FM 3-0).

3-2. Command is an individual and personal function. It blends imaginative problem solving, motivational and communications skills, and a thorough understanding of the dynamics of operations. Command during operations requires understanding the complex, dynamic relationships among friendly forces, enemies, and other aspects of the operational environment (OE). This understanding helps commanders visualize and describe their intent and develop focused planning guidance. Command is a specific and legal leadership responsibility unique to the military (see FM 6-22).

CONTROL

3-3. Control is the regulation of forces and WFFs to accomplish the mission in accordance with the commander's intent (FM 3-0). Control is fundamental to directing operations. Commanders and staff both exercise control. Commanders and staffs must understand the science of control to overcome the physical and procedural constraints under which units operate. Control also requires a realistic appreciation for time-distance factors and the time required to initiate certain actions. It demands understanding those aspects of operations that can be analyzed and measured. It relies on objectivity, facts, empirical methods, and analysis.

3-4. Control of sustainment spans the strategic to tactical level. For example, the control of the flow of sustainment across all levels of war is crucial for supporting operations. It demands an understanding of sustainment functions and related systems that support all aspects of the distribution process. It also requires the availability of organizations, centers, and activities designed with the mission to control sustainment. As a result; Movement Control Battalions (MCB), DMCs, materiel management sections, and support operations (SPO) sections within sustainment commands are responsible this control. Brigades and battalions are primarily responsible for the control and coordination of distribution operations. These activities rely on a number of information systems to help control sustainment. Appendix A contains details on the various information systems.

BATTLE COMMAND AND SUSTAINMENT

3-5. Battle command is the art and science of understanding, visualizing, describing, directing, leading, and assessing forces to impose the commander's will on a hostile, thinking, and adaptive enemy. Battle command applies leadership to translate decisions into actions—by synchronizing forces and WFFs in time, space, and purpose—to accomplish missions. Battle command is guided by the commander's professional judgment gained from experience, knowledge, education, intelligence, assessment skills, intuition, and leadership. Figure 3-1 diagrams battle command.

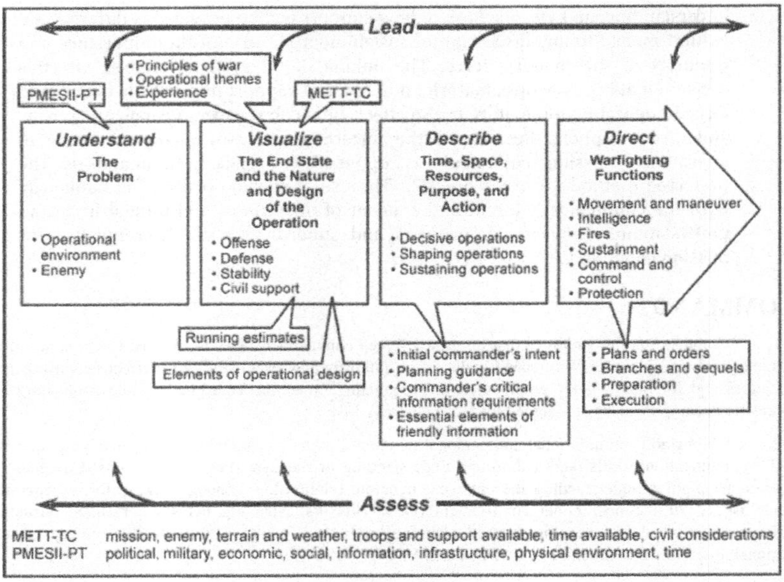

Figure 3-1. Battle Command

3-6. The sustainment commanders must have broad perspective, understanding, and knowledge of sustainment activities throughout the operational area. They must share the visualization of the operational commander and then how to employ all elements of sustainment capabilities at their disposal in support of the operation. Then they must describe and direct how these capabilities are provided. The elements of battle command from the perspective of the sustainment commander are discussed below.

UNDERSTAND

3-7. Understanding is fundamental to battle command. Sustainment commanders must first understand the supported commanders' intent and concept of operations. They understand how and what the supported commander thinks. They specifically must understand the supported commander's intent and concept of the operations then track developments and adjust plans as the operations unfold. Sustainment commanders must understand processes and procedures for the provision of sustainment, in relation to the operational environment and the resources available to them. They must understand the relationship between each of the WFFs and how sustainment impacts each. Sustainment commanders must also understand the flow of sustainment and the critical decision points at which they can effect or adjust resources based on changing mission requirements.

3-8. Understanding changes as operations progress. Understanding the changes to the operation allows the commander to choose and exploit METT-TC factors that best support the mission. Sustainment commanders build upon their understanding by collecting, storing, and sorting through information that impacts the operation. As a result, the sustainment estimate and commanders' understanding have to be reviewed and re-evaluated throughout an operation. They use a variety of tools, methods, and resources to increase their understanding. A few are discussed below.

Relevant Information (RI)

3-9. RI is defined as information of importance to commanders and staffs in the exercise of C2 (FM 3-0). RI provides the answers commanders and staffs need to conduct operations successfully. Effective information management helps staffs collect and store information commanders need for better understanding and thus knowledge of the mission, task, or operation. For the sustainment commander, RI drives how he/she visualizes the concept of support. It aides the sustainment commander in determining what, when, and where support is needed. It helps commanders anticipate requirements and prioritize support for current and future operations. It also helps in understanding potential problem areas that may need resolving before impacting support.

Intelligence

3-10. Intelligence Preparation of the Battlefield (IPB) is a continuous staff planning activity undertaken by the entire staff to understand the operational environment and options it presents to friendly and threat forces. It is a systematic process of analyzing and visualizing the operational environment in a specific geographic area and for each mission. By applying IPB, commanders gain the information necessary to selectively apply and maximize combat power at critical points in time and space. Understanding intelligence data is critical to sustainment operations. Analysis of intelligence information may help commanders avoid potential enemy activity and threats. Sustainment Soldiers in many ways (such as convoy operations and support to stability operations) become a valuable source for collecting intelligence data which must be processed and passed through intelligence channels.

Liasion

3-11. Liaison is that contact or intercommunication maintained between elements of military forces or other agencies to ensure mutual understanding and unity of purpose and action (JP 3-08). The Liaison Officer (LNO) is the personal and official representative of the sending organization commander and should be authorized direct face-to-face liaison with the supported commander. LNOs must have the commander's full confidence and the necessary rank and experience for the mission. Using an LNO conserves manpower while guaranteeing the consistent, accurate flow of information, coordination, advice, and assistance.

3-12. LNOs are essential for the sustainment mission for several reasons. Through monitoring of the supported command's mission, the sustainment LNOs provide quick information on mission changes thus enabling responsive adjustments in support of the operation. The LNO enables sustainment command staffs and supported command staffs in their planning and coordination, thereby assuring unity of effort. The LNO is an important advisor to the supported commander aiding in the employment of sustainment assets.

Command Visits

3-13. Another technique used to facilitate understanding is command and staff visits with supported commands. It enables commanders to determine the implications of what is happening (situational awareness) and anticipate what may happen (commander's visualization). It also establishes character, presence, and intellect (attributes of leadership) and instills competence. It enables commanders to see firsthand, the operational environment and the supported commander's mission. As a result, discussion, comparison of views, and continuous study facilitates situational understanding. Sustainment commanders and their staffs obtain a better understanding of the requirements of supported units and the operational environment in which they operate. To maintain situational understanding, commanders talk with their peers, subordinates, superiors, and with their staffs, and with community and civilian agency leaders. This assures sustainment commanders are better able to integrate sustainment into operations, anticipate support requirements, and provide responsive and continuous support.

Information Systems

3-14. An information system is equipment and facilities that collect, process, store, display, and disseminate information. This includes computers—hardware and software—and communications as well as policies and procedures for their use (see FM 3-0). The commander uses these systems to collect information. Having access to these systems gives the commander a common operational picture (COP). The COP is a single display of RI within a commander's area of interest tailored to the user's requirements and based on common data and information shared by more than one command (FM 3-0).

3-15. For example, BCS3 provides support to commodity tracking, convoy operations, and the RSO&I process. The system also produces user-defined COP. BCS3 fuses sustainment information from numerous sources into one user-defined; mission focused and tailored, map-centric visual display. It provides graphics that are scalable to the display requirements of tactical, operational, and strategic users' needs. It exchanges C2 information with Army Battle Command System (ABCS) and joint systems. It also integrates logistics and personnel services information required to manage sustainment operations.

3-16. Another example is the Movement Tracking System (MTS). MTS allows sustainment organizations to track, in real time, truck locations, communicate with the drivers, redirect cargo as required, and provide In-Transit Visibility. This capability also provides the ability to avoid identified hazards, inform operators of unit location changes, and provide Traffic Regulation and Control. MTS Control Stations are located in various sustainment C2 and staff sections to provide overview of convoy operations within a specified area. A discussion of MTS and other sustainment related information systems is in Appendix A.

VISUALIZE

3-17. Visualization follows the commanders understanding. Commander's visualization is the mental process of developing situational understanding, determining a desired end state, and envisioning the broad sequence of events by which the force will achieve that end state (see FM 3-0). Understanding helps the commander to pull all of the pieces of the puzzle together to build the picture in his/her mind. The sustainment commander's visualization requires him/her to picture current and future operations and how to employ sustainment assets and resources in support. His/Her visualization takes into account several factors such as METT-TC, defining the end state, and determining the most effective method for employing availability of sustainment resources. He/She must also consider the principles of sustainment and the integration of the WFFs.

End State

3-18. The end state is a set of required conditions that defines achievement of the commander's objectives (JP 1-02). In terms of sustainment, determining the end state may be complex. For the sustainment commander, achieving the desired end state involves determining the most effective means for getting the supported commander what he/she needs, when he/she needs it, and where he/she needs it to conduct full spectrum operations.

3-19. As defined, sustainment is the provision of logistics, personnel services, and health services support to maintain the force until mission accomplishment. Commanders at each level (strategic, operational, and tactical) have different focuses and perhaps different definitions of end state.

3-20. At the strategic level, the sustainment commanders' focus is on force readiness. The end state is the ability of a combat force to mobilize, deploy, sustain, redeploy, and reset. The key sustainment end state is continuous cycle of ensuring units are equipped, manned, and healthy to conduct operations globally. At the strategic level, the sustainment commander's visualization may include, but not be limited to, what budget requirements are needed to fund readiness initiatives, and how to modernize forces to make them more combat effective while minimizing deployment resources. It may include things such as what forces are needed to support the GCC's operation based on METT-TC and priority for employing sustainment forces to support theater operations. Another key end state may be determining what generating forces support is needed for the operating force.

3-21. At the operational level, the end state is more narrowly focused. While readiness is a critical factor, the sustainment end state at this level may be the distribution of sustainment to support the GCC mission. It is focused on continuity of support and how best to enable the operational reach of Army forces. It may also be how to be more responsive to the needs of the commander.

3-22. At the tactical level, the end state is the uninterrupted provision of sustainment to all units to support continuous operations in an assigned area. As a result, visualizing an end state is a continuous process and requires continuous monitoring of the situation. Commanders may make adjustments as the situation may rapidly change to any combination of offense, defense, and stability support.

Mission, Enemy, Terrain and Weather, Troops and Support Available, Time Available, and Civil Considerations (METT-TC)

3-23. The assignment of a mission provides the focus for developing the commander's visualization. Commanders use METT-TC as a means for identifying mission variables.

Mission

3-24. The mission is the task, together with the purpose, that clearly indicates the action to be taken and the reason therefore (JP 1-02). Commanders analyze a mission in terms of specified tasks, implied tasks, and the commander's intent two echelons up. They consider the missions of adjacent units to understand their relative contributions to the decisive operation.

3-25. Sustainment commanders must understand the supported commander's mission. The supported commander establishes the priority of support. Since sustainment is generally provided to a designated area of operations, commanders carefully assess the operational mission to determine the types of units operating in the area and their sustainment needs. Results of that analysis yield the essential tasks that—with the purpose of the operation—clearly specify the sustainment actions required which then become the sustainment unit's mission. The sustainment commander and staff work closely with the operational staffs to ensure the integration of sustainment with the operations and mission plans.

Enemy

3-26. The enemy may consider such sustainment operations, as convoys and medical evacuations, as relatively soft targets to attack. In the current operational environment, the enemy has used IEDs and ambushes on convoys as one of the methods to disrupt sustainment operations. To reduce this risk, sustainment commanders may develop, alter, and/or improvise plans and actions for avoiding potential sites for attacks. Understanding when and how the enemy is most likely to attack require detailed, timely, and accurate information. Effective intelligence, surveillance, and reconnaissance information is important for identifying threat capabilities and vulnerabilities.

Terrain and Weather

3-27. Terrain includes both natural and man-made features such as rivers, mountains, cities, airfields, and bridges. Weather includes atmospheric conditions such as excessive heat, cold, rain, snow, and a variety of storms. Terrain and weather significantly impact sustainment. It influences the sustainment commander's decision and visualization for supporting operations. Sustainment commanders visualize the advantages and disadvantages afforded by terrain and weather. Natural terrain features may help conceal sustainment forces or operations. On the other hand, it may provide opportunities for enemy attack. Urban areas may provide more access to contract capabilities, but may also serve as bottlenecks for convoys or impede medical support. Weather likewise has advantages and disadvantages. For example, cloud cover may conceal sustainment operations from aerial attack while it may also hinder aerial delivery of supplies or MEDEVAC operations.

3-28. Terrain and weather also influences the type of sustainment provided. For example, urban operations may require increased quantities of small arms and crew served ammunition versus tank or artillery ammunition for open terrain. Weather factors such as heat or cold will increase the demand on supplies such as water or cold weather equipment. It may also increase incidences of disease and non-battle injuries (DNBI).

Troops and Support Available

3-29. Troops and support available is the number, type, capabilities, and condition of available friendly troops. These include resources from joint, interagency, multinational, host nation, commercial, and private organizations. It also includes support provided by civilians (FM 3-0). Troops and support available falls largely within the sustainment area and encompasses much of the sustainment commander's visualization. Sustainment staffs track readiness including training, maintenance, logistics, health and welfare, and morale. They provide commanders with the required information to enable him/her to make decisions about his/her ability to execute assigned missions. Sustainment commands and staffs are responsible for ensuring required resources are available and replacements are requested to fill shortfalls.

Time Available

3-30. Time is critical to the responsiveness of all operations. Sustainment commanders must understand the time sensitive nature of operations and maximize all available time to get commanders what they need when they need it. A key consideration for sustainment commander's visualization is the time it will take to get resources to supported commanders. There are numerous techniques sustainment commanders may use to resolve timing issues. One solution may be positioning support in proximity of the operations. Another solution is the ability to anticipate support requirements and initiate processes and procedures to begin the flow of support through the sustainment chain. In certain unique circumstances, the commander may have to improvise or take risk assuring critical support is provided. He/She may also direct by passing support nodes or aerial delivery or other means.

Civil Considerations

3-31. Understanding the operational environment requires understanding civil considerations. Civil considerations reflect man-made infrastructure, civilian institutions, attitudes, and activities. The civilian leaders, populations, and organizations within an area of operations influence the conduct of military operations. In instances where stability and offensive operations are concurrent, sustainment commanders may be required to provide support to civilian populations in addition to ongoing military operations, until other agencies or HNS is available. This may be particularly true during resettlement operations. Supporting such operations places heavy demands on sustainment forces and activities. Military movements, supply distribution, contracting, and other sustainment activities may be strained.

3-32. Commanders should avoid providing support to civilian populations that compete with economic factors in the community. For example, providing services to a population that take away from businesses or jobs may have an adverse effect on the local economy. Correspondingly, contracting for these services contributes to fostering local support and may boost the local economy. The sustainment preparation of the operational environment, discussed in Chapter 4, is one technique commanders may use to visualize civil considerations in support of operations.

DESCRIBE

3-33. The visualization process results in commanders describing to their staffs and subordinates the shared understanding of the mission and intent. Commanders ensure subordinates understand the visualization well enough to begin planning. Commanders describe their visualization in doctrinal terms, refining and clarifying it as circumstances require. Commanders express their initial visualization in terms of:

- **Initial commander's intent**. Commanders summarize their visualization in their initial intent statement. The purpose of the initial commander's intent is to facilitate planning while focusing on the overall operations process. The sustainment commander's intent should reflect his/her visualization for supporting the operational commander. His/Her intent must integrate elements of the operational commander's intent to ensure synchronization and unity of effort. The sustainment staffs must analyze the commander's intent to ensure supportability of the operation.
- **Planning guidance.** Planning guidance conveys the essence of the commander's visualization. It broadly describes when, where, and how the commander intends to employ combat power to accomplish the mission. Sustainment commander's guidance conveys his/her vision for sustaining combat power. His/Her guidance may include such factors as the placements of sustainment assets to best provide responsive support. It may include guidance for supply rates or evacuation requirements. His/Her guidance may also establish priorities of support based on the missions within his/her designated support area. His/Her planning guidance ensures staffs understand the broad outline of his/her visualization while allowing the latitude necessary to explore different options.
- **Commander's Critical Information Requirements (CCIR).** A commander's critical information requirement is an information requirement identified by the commander as being critical to facilitating timely decision making. One of the staff's priorities is to provide the commander with answers to CCIR. Some examples of CCIR for sustainment commanders may be: What are the consumption rates for various classes of supply? What and where are those supplies in the distribution pipeline? Where are the most likely casualties to occur and are there assets available to evacuate them? What type and where are personnel replacements needed? What is the maintenance status of critical combat equipment? While most staffs provide RI, a good staff expertly distills that information. It identifies answers to CCIR and gets them immediately to the commander. It also identifies vital information that does not answer a CCIR, but that the commander nonetheless needs to know. The two key elements are friendly force information requirements and priority intelligence requirements (JP 3-0).
- **Essential elements of friendly information (EEFI).** An *essential element of friendly information* is a critical aspect of a friendly operation that, if known by the enemy, would subsequently compromise, lead to failure, or limit success of the operation, and therefore should be protected from enemy detection. An EEFI establishes an element of information to protect rather than one to collect. For sustainment, a few examples of EEFI may include readiness status of units or critical personnel, equipment, and/or maintenance shortfalls. Other factors may be supply routes or schedules for resupply operations and locations of essential stocks or resources.

DIRECT

3-34. Commanders direct all aspects of operations. This direction takes different forms throughout the operations process. Commanders make decisions and direct actions based on their situational understanding, which they maintain by continuous assessment. They use control measures to focus the operation on the desired end state. Sustainment commanders direct operations by:

- Assigning and adjusting missions, tasks, task organization, and control measures based on the concept of operations of the supported commander and his/her priorities.
- Positioning units to maximize support, anticipate combat operations and changes to combat intensity, or create or preserve maneuver options.

- Positioning key leaders to ensure observation and supervision at critical times and places.
- Adjusting support operations to execute changed priorities based on exploiting opportunities or avoiding threats.

3-35. Sustainment commanders direct forces by issuing orders to subordinates. The orders issued may be verbal, written, or electronically produced using matrices or overlays. These orders may be of several types (see FM 5-0).

- An *operation order (OPORD)* is a directive issued by a commander to subordinate commanders for the purpose of effecting the coordinated execution of an operation (FM 5-0).
- The *warning order (WARNO)* is a preliminary notice of an order or action, which is to follow. WARNOs help subordinate units and staffs prepare for new missions.
- A *fragmentary order (FRAGO)* is an abbreviated form of an operation order (verbal, written, or digital) usually issued on a day-to-day basis that eliminates the need for restating information contained in a basic operation order.

STAFF

3-36. Staffs assist the commander in providing control over and executing timely decisions for operations. Commanders and staffs are continually alert for opportunities to streamline cumbersome or time-consuming procedures. They provide RI to help commanders achieve situational understanding. One piece of information alone may not be significant; however, when combined with other information from the COP, it may allow the commander to formulate an accurate visualization and make an appropriate decision.

SUSTAINMENT CELL ROLES

3-37. Modular force theater army, corps, and division headquarters have been realigned in accordance with the WFFs of Movement and Maneuver, Fires, Intelligence, Command and Control, Protection, and Sustainment. The WFF Cell – Sustainment combines many of the functions formerly found in G1, G4, G8, and Surgeon staffs, and the Engineer Coordinator (ENCOORD). These functions are now organized into a G1 Division, G4 Division, G8, Surgeon, and Engineer Division (see Figure 3-2).

SUSTAINMENT STAFF COORDINATION

3-38. The sustainment staffs are responsible for providing staff support activities for the commander. The sustainment staff integrator monitors and coordinates sustainment functions between the sustainment staffs and other WFF staffs and advises the commander on force readiness. The sustainment staffs also coordinate with the corresponding sustainment commands and specialized functional centers (HRSC and FMC) for oversight of operations.

G-1 STAFF

3-39. The G-1 staff is an element of Theater Army sustainment HQ and staff section of corps or division. At each command level, the G-1 is the senior HR advisor to the commander. The G-1's mission is to ensure HR readiness and plan HR support in compliance with the commander's priorities, intent, and policies. The HR mission is accomplished through a combination of direct communication with the USAHRC, G-1 and S-1 staffs at Theater Army, corps and division, and TSCs and HRSC.

3-40. The G-1 focus is establishing HR policy, priorities, and providing guidance and oversight for the HR functions. These functions include man the force, provide HR services, coordinate personnel support, and conduct HR planning and staff operations.

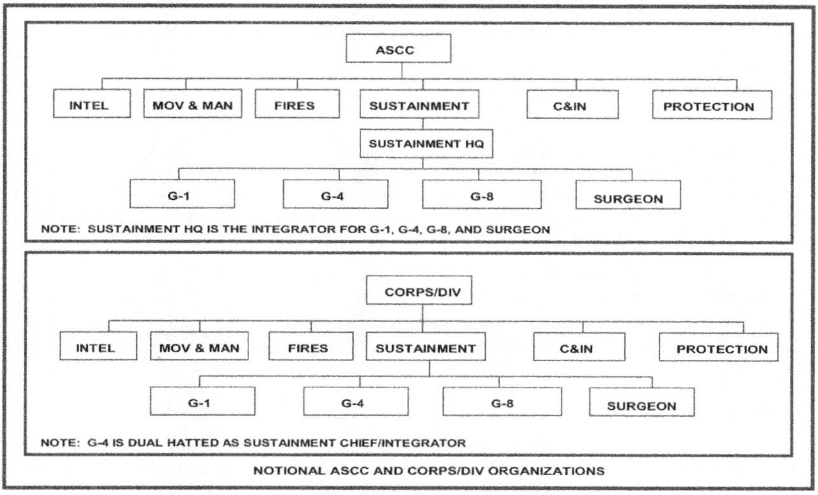

Figure 3-2. Notional Theater Army and Division or Corps Sustainment Cells

3-41. The G-1 coordinates through the TSC to the HRSC for personnel accountability (PA), reception, replacement, and return to duty, rest and recuperation, redeployment, postal, and casualty operations. The G-1 also advises the commander on the allocation and employment of HR support and units within the AOR (see FM 1-0). Other G-1 responsibilities include, but are not limited to:

- Coordinating HR functions across the entire JOA when serving as a Joint Task Force (JTF) Combined Force Land Component Commander/Joint force Land Component Commander (CFLCC/JFLCC) C-1/J-1 (see JP 1-0).
- Planning and integrating civilian personnel HR support.
- Incorporate unique Reserve Component programs and policies into theater HR programs.

G-4 STAFF

3-42. The G-4 is the principal staff officer responsible for monitoring and advising on logistics within the Theater Army, corps, or division commands. At the corps and division, the G-4 also serves as the sustainment chief/integrator for the commander. The G-4 staff develops the logistics plan in support of the operational plan. The staff provides recommendations on a variety of command priorities including force structure, HNS, contracting support, materiel management, and movement control.

3-43. The G-4 staff may include divisions, branches, and specialized sections for supporting various types of operations. In addition, it may have joint and multinational capabilities for supporting requests for logistics support to joint and MNFs, U.S. Agencies, NGOs, and IGOs in the theater. These logistic requests are coordinated through the G-9 (Assistant Chief of Staff for Civil-Military Operations) or the civil-military operations center.__The staff may also serve as the focal point for the coordinating LOGCAP and other contracting support.

G-8 STAFF

3-44. The G-8 is the senior FM advisor to the commander at ASCC, corps, and division levels. The staff's mission is to fund the force through the coordination and synchronization of resources against unit requirements. It provides advice and guidance concerning resources commanders and staff. It also synchronizes RM operations and performs management requirements as determined by the commander.

3-45. The G-8 performs the following functions:

- Acquires, manages, and distributes funds.
- Manages DOD resources.
- Plans, programs, and develops budget.
- Coordinates and supports accounting.
- Supervises and provides leadership to the G-8 staff.
- Estimates, tracks, and reports costs for specific operations to support requests to the U.S. Congress for appropriation.

3-46. The Army may be appointed the lead Service responsible for common FM support. If so, Army G-8 staffs ensure that RM, banking, and currency support are provided for joint interagency and multinational operations. The Army also provides financial analysis and recommendations to joint forces for the most efficient use of fiscal resources. FM provides the funding essential to support contracting requirements and accomplish joint programs.

SURGEON

3-47. A command surgeon is designated for all command levels. This medical officer is a special staff officer charged with advising on the AHS mission. The duties and responsibilities of command surgeons for HSS include: Advise the commander on the health of the command; plan and coordinate AHS support for units in theater; establish polices for care of non-U.S personnel; maintain HSS situational awareness; monitor troop strength of medical personnel; monitor medical logistics and blood management operations; and monitor medical regulating and patient tracking operations.

3-48. The duties of command surgeons for FHP include: coordinate for veterinary support for food safety, animal care, and veterinary preventive medicine; plan and implement FHP operations to counter health threats; recommend combat and operational stress control, behavioral health, and substance abuse control programs; advise commanders on FHP CBRN defensive actions; ensure the provision of dental services; and ensure a medical laboratory capability for the identification and confirmation of the use of suspect biological warfare and chemical agents.

3-49. The Theater Army surgeon ensures that all AHS support functions are considered and included in OPLANs and OPORDs. The surgeon is a full-time special staff officer answering directly to the Theater Army commander on matters that pertain to the health of the command. He/She coordinates for AHS support for both HSS and FHP. The Theater Army surgeon maintains a technical relationship with the MEDCOM (DS) commander and helps establish medical policy for the theater.

3-50. The corps surgeon is solely responsible for planning, coordinating, and synchronizing the AHS effort within the Corps AO. The corps surgeon maintains a technical relationship with the MEDCOM (DS) commander and helps establish medical policy for the AO.

CHAPLAIN

3-51. The chaplain is a member of the commander's personal staff. The chaplain is responsible for religious support operations. The chaplain advises the commander on matters of religion, morals, and morale as affected by religion and on the impact of indigenous religions on military operations. No coordinating staff officer exercises responsibility over the chaplain.

JUDGE ADVOCATE GENERAL

3-52. The staff judge advocate (SJA) is a member of the commander's personal staff. The HQ legal team participates in actions related to C2 of its subordinates. Command and staff functions include advice to commanders, staffs, and Soldiers on the legal aspects of command authority, command discipline, applying force, and the law of war.

3-53. Legal support to sustainment activities includes negotiating acquisition and cross-servicing agreements, SOFAs with host nations, combat contingency contracting, fiscal law, processing claims arising in an operational environment, and environmental law.

SUMMARY

3-54. C2 is the exercise of authority and direction by a properly designated commander over assigned and attached forces. Sustainment commanders and staffs use C2 of sustainment units to provide support to operations. Sustainment commanders apply the principles of battle command-- understanding, visualizing, describing, directing, and assessing, to lead sustainment forces. They also use the variables of METT-TC to assist in visualizing support. The sustainment staffs of operational HQ conduct vertical and horizontal staff coordination. Sustainment staffs in Theater Army, corps, and divisions assist the commander in the operations process for providing support. Sustainment staffs at Theater Army are organized under a sustainment HQ element. At corps and division, the G-4 serves the dual role as G-4 and sustainment chief/integrator for the sustainment staff elements. A sustainment staff consists of the G-1, G-4, G-8, and representatives from the Surgeon, and ENCOORD. The chaplain and judge advocate general staff serves as personal staff to the commander.

Chapter 4

Integrating Sustainment into Operations

Integration is a key principle of sustainment. Effective integration of sustainment sets the conditions to ensure mission success and extend strategic and operational reach. Integration begins with the operations process—planning, preparation, execution, and continuous assessment. It is conducted simultaneously and in synchronization with the operations plan development. Sustainment must be integrated across each level of war and with joint and multinational operations. This chapter covers how sustainment is integrated into the operations process to support full spectrum operations—offense, defense, stability, and civil support operations.

INTEGRATING SUSTAINMENT INTO THE OPERATIONS PROCESS

4-1. *The operations process consists of the major C2 activities performed during operations: planning, preparing, executing, and continuously assessing the operations (FM 3-0).* Integrating sustainment with the operations process across each level of war is vital for ensuring the synchronization of sustainment with the WFFs and unity of effort during operations. Integrating sustainment with joint and multinational operations allow forces to conduct operations using mutual support capabilities while reducing redundancy and competition for limited resources. Commanders and staff s at every level must make all efforts to integrate sustainment with the operations process. Figure 4-1 illustrates the operations process.

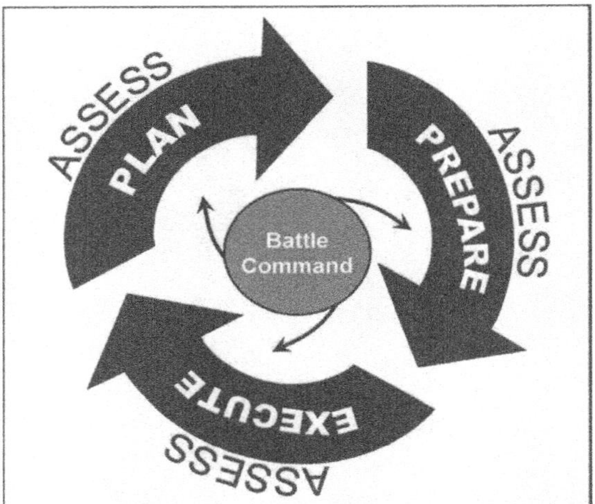

Figure 4-1. The Operations Process

PLANNING THE SUSTAINMENT OF OPERATIONS

4-2. Chapter 3 covered many of the C2 planning functions (such as battle command, the determination of end state, and sustainment staff roles). This chapter will focus on more specific tools and planning considerations sustainment commanders and staffs use in planning for sustainment of full spectrum operations.

4-3. Planning begins with analysis and assessment of the conditions in the operational environment with emphasis on the enemy. It involves understanding and framing the problem and envisioning the set of conditions that represent the desired end state (FM 3-0). Sustainment planning indirectly focuses on the enemy but more specifically on sustaining friendly forces to the degree that the Army as a whole accomplishes the desired end state. There are several tools available for conducting course of action analysis. We will highlight a couple of them below.

SUSTAINMENT PREPARATION OF THE OPERATIONAL ENVIRONMENT

4-4. Sustainment preparation of the operational environment is the analysis to determine infrastructure, environmental, or resources in the operational environment that will optimize or adversely impact friendly forces means for supporting and sustaining the commander's operations plan. The sustainment preparations of the operational environment assist planning staffs to refine the sustainment estimate and concept of support. It identifies friendly resources (HNS, contractable, or accessible assets) or environmental factors (endemic diseases, climate) that impact sustainment.

4-5. Some of the factors considered (not all inclusive) are as follows:

- Geography. Information on climate, terrain, and endemic diseases in the AO to determine when and what types of equipment are needed. For example, water information determines the need for such things as early deployment of well-digging assets and water production and distribution units.
- Supplies and Services. Information on the availability of supplies and services readily available in the AO. Supplies (such as subsistence items, bulk petroleum, and barrier materials) are the most common. Common services consist of bath and laundry, sanitation services, and water purification.
- Facilities. Information on the availability of warehousing, cold-storage facilities, production and manufacturing plants, reservoirs, administrative facilities, hospitals, sanitation capabilities, and hotels.
- Transportation. Information on road and rail networks, inland waterways, airfields, truck availability, bridges, ports, cargo handlers, petroleum pipelines, materials handling equipment (MHE), traffic flow, choke points, and control problems.
- Maintenance. Availability of host nation maintenance capabilities.
- General Skills. Information on the general skills such as translators and skilled and unskilled laborers.

OPERATIONS LOGISTICS (OPLOG) PLANNER

4-6. The OPLOG Planner is a web-based interactive tool that assists commanders and staff from strategic through operational levels in developing a logistics estimate. It is designed to support operations typically associated with multi-phase operations plans and orders. The OPLOG Planner enables staffs to develop estimated mission requirements for supply Class I, Class II, Class III(P), Class IV, Class VI, Class VII, Class X including water, ice, and mail. The tool uses the latest Army approved planning rates and modular force structures. It is updated at least annually to stay current with force structure and rate changes.

4-7. OPLOG Planner allows planners to build multiple task organizations from a preloaded list of units and equipment or from custom built units that are generated or imported. Each task organization is assigned a consumption parameter set that establishes the rates, climate, Joint Phases (Deter, Seize Initiative, and Dominate), and Army full spectrum operations (Offense, Defense, and Stability). The planners have the option of using predefined default planning rates or customizing rates based on what a unit is experiencing.

4-8. OPLOG Planner generates the logistics supply requirements which can be viewed by the *Entire Operation*, each *Phase of the Operation*, each *Task Organization*, each *Unit*, or each *Unit's Equipment*. Logistics requirement reports can be printed or exported to automated spread sheets for further analysis or saved for recall to be used in course of action analysis.

PLANNING ARMY HEALTH SERVICES

4-9. Medical planners determine the capabilities and assets needed to support the mission. To ensure effective and efficient support, medical plans adhere to the principles of AHS support (as outlined in Chapter 5), the commander's planning guidance, medical intelligence related to the operational area, and other planning considerations.

4-10. The theater evacuation policy, health threat, troop strength or size of the supported population, and the type, intensity and duration of the operation are some of the factors considered for determining medical requirements. The medical staff estimates and patient estimates are also developed during planning. The patient estimate is derived from the casualty estimate prepared by the G-1.

4-11. In-depth analysis is critical at every level of the operation to ensure the flexibility to quickly react to changes in the mission and continue to provide the required support. The observations of commanders, disease and nonbattle injury rates, and running estimates are the primary means of assessing an operation to ensure that the concept of operations, mission, and commander's intent are met. These factors and continuous analysis help to make certain that once developed, the plan includes the right number and combination of medical assets to support the operation. See FM 8-55 for additional information.

4-12. Medical and nonmedical automated information systems are used to plan AHS mission. Medical commanders must know the complete COP which includes situational awareness of three areas—(1) tactical (via FBCB2), (2) medical (via MC4/DHIMS), and (3) sustainment (via GCSS-Army). The commanders and medical planners must maintain situational awareness, in-transit visibility and tracking of patients and equipment, and a common operational picture of the AO. This information is obtained through various plans, reports, and information systems available to commanders and planners to facilitate the decision making process. These and other medical systems are discussed in Appendix A.

PLANNING CONSIDERATIONS FOR FULL SPECTRUM OPERATIONS

4-13. Army forces plan offensive, defensive, and stability or civil support operations simultaneously as part of an interdependent joint force. The proportion and role of offensive, defensive, and stability or civil support tasks are based on the nature of the operations, tactics used, or the environment. While full spectrum operations may occur simultaneously versus sequentially, they have slightly different planning requirements. Figure 4-2 illustrates the element of full spectrum operations.

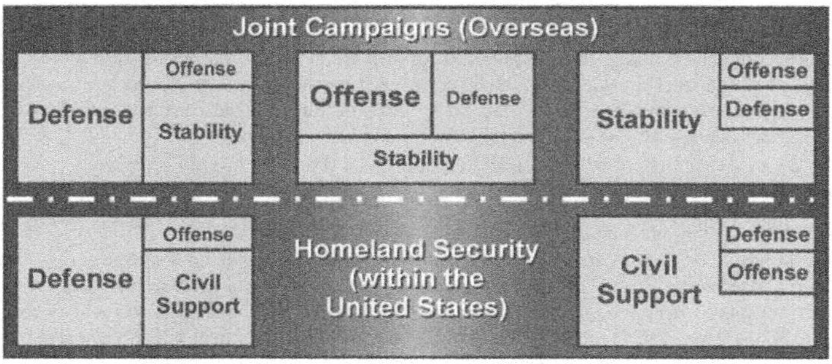

Figure 4-2 .Elements of Full Spectrum Operations

OFFENSIVE OPERATIONS

4-14. *Offensive operations are defined as combat operations conducted to defeat and destroy enemy forces and seize terrain, resources, and population centers. They impose the commander's will on the enemy (see FM 3-0).* Sustainment of offensive operations is high in intensity. Commanders and staffs plan for increased requirements and demands on sustainment. Sustainment planners work closely with other WFF staffs to determine the scope of the operations and develop estimates for quantity and types of support required. They anticipate where the greatest need might occur and develop a priority of support. Sustainment planners may consider positioning sustainment units in close proximity to operations to reduce response times for critical support. They also consider alternative methods for delivering sustainment in emergency situations.

4-15. To maintain momentum and freedom of action, coordination between staff planners must be continuous. During offensive operations, certain requirements present special challenges. The most important materiel is typically fuel (Class III Bulk) and ammunition (Class V), Class VII, movement control, and medical evacuation. Based on planning assessments, sustainment commanders direct the movement of these and other support to meet anticipated requirements.

4-16. Another challenge in planning for and sustaining an offensive operation is the lengthened lines of communication (LOCs). Widely dispersed forces, longer LOCs, and congested road networks increase stress on transportation systems. As a result, a combination of ground and aerial delivery maybe planned to accommodate the distribution. Distribution managers and movement control units synchronize movement plans and priorities according to the commander's priority of support. Distribution must be closely coordinated and tracked to ensure delivery of essential support. The routing function of movement control becomes an essential process for coordinating and directing movements on main supply routes or alternate supply routes, and regulating movement on LOCs to prevent conflict and congestion.

4-17. Higher casualty rates associated with offensive operations increase the requirement of medical resources. Plans to position medical support close to operations to facilitate treatment and evacuation are considered. If increased casualty rates overwhelm medical resources, nonmedical transportation assets may be needed for evacuation. Another planning consideration may be moving combat and operational stress control teams to support combat stress casualties following operations.

4-18. Higher casualty rates also increase the emphasis on personnel accountability, casualty reports, and replacement operations. G-1s and S-1s plan for accurate tracking of casualties and replacements through coordination with Casualty Liaison Teams (CLTs) and the HRSC.

4-19. Plans should also provide for religious support, which may become critical during offensive operations. Religious support through counseling and appropriate worship can help reduce combat stress, increase unit cohesion, and enhance performance.

4-20. Using contractors during the sustainment of offensive operations entails great risk and raises significant practical and legal considerations. However when necessary, the force commander may be willing to accept this risk and use contractors in forward areas. Commanders should seek counsel from their judge advocates when considering the use of contractors during offensive operations.

DEFENSIVE OPERATIONS

4-21. *Defensive operations defeat an enemy attack, gain time, economize forces, and develop conditions favorable for offensive or stability operations (FM 3-0).* The commander positions assets so they can support the forces in the defense and survive. Requirements for sustainment of forces in the defense depend on the type of defense. For example, increased quantities of ammunition and decreased quantities of fuel characterize most defensive operations. However, in a mobile defense, fuel usage may be a critical factor. Conversely, in a fixed defensive position, fuel requirements are lower.

4-22. Typically, defensive operations require more centralized control. Movements of supplies, replenishment, and troops within the AO have to be closely and continuously coordinated, controlled, and monitored. In retrograde operations (a type of defense) special care is necessary to assure that combat units receive necessary support across the depth of the AO and that the sustainment unit and stocks are not lost as the unit moves away from enemy activity.

4-23. Distribution managers direct resupply forecasted items to designated units. Increases in items such as barrier and construction materials should be pushed to designated collection points for unit retrieval whenever possible. Planners should consider the impact of increased ammunition expenditures on available transportation assets.

4-24. The task of medical units is to triage casualties, treat, and return to duty or resuscitate and stabilize for evacuation to the next higher level of medical care or out of the theater of operations. MTFs and other AHS support assets should be placed within supporting distance of maneuver forces but not close enough to impede ongoing combat operations.

STABILITY OPERATIONS

4-25. *Stability operations encompass various military missions, tasks, and activities conducted outside the United States in coordination with other instruments of national power to maintain or reestablish a safe and secure environment, provide essential governmental services, emergency infrastructure reconstruction, and humanitarian relief (FM 3-0).* Although Army forces focus on achieving the military end state, they ultimately need to create conditions where the other instruments of national power are preeminent. Sustainment of stability operations often involves supporting U.S. and MNFs in a wide range of missions and tasks. The tasks most impacted by sustainment are briefly discussed below.

Establish Civil Security

4-26. *Civil security involves providing for the safety of the host nation and its population, including protection from internal and external threats (FM 3-0).* Sustainment staffs must consider plans to support I/R operations. Sustainment may be provided to these operations until HNS, NGO, and OGOs are available.

4-27. Sustainment planners must address the sustainment of I/R operations. Sustainment to I/R involves all of the sustainment functions. Planners should consider general engineering support horizontal and vertical construction of detention centers, as well as repair and maintenance of the infrastructure (see FM 3-34.400). A more detailed discussion of sustainment support of I/R and detainee operations will be provided later in this chapter.

Establish Civil Control

4-28. Legal staffs should plan for supporting civil control operations. Civil control centers on rule of law, supported by efforts to rebuild the host nation judiciary and corrections systems. It encompasses the key institutions necessary for a functioning justice system, including police, investigative services, prosecutorial arm, and public defense. If transitional military authority is instituted, intervening forces will likely carry out judicial and correctional functions.

Restore Essential Services

4-29. Efforts to restore essential services involve developing host nation capacity to operate, maintain, and improve those services. At the tactical or local level, sustainment in support of civil authorities will restore essential civil services as defined in terms of immediate humanitarian needs (such as providing food, water, shelter, and medical and public health support) necessary to sustain the population until local civil services are restored. Other sustainment tasks associated with restoration of services include support to dislocated civilians and demining operations.

Support to Economic and Infrastructure Development

4-30. The role of sustainment in supporting economic stabilization and infrastructure development is significant, especially at the local level. Here the emphasis is on generating employment opportunities, infusing monetary resources into the local economy, stimulating market activity, fostering recovery through economic development, and supporting the restoration of physical infrastructure. Drawing on local goods, services, and labor through contracting, presents the force an opportunity to infuse cash into the local economy, which in turn stimulates market activity.

4-31. Restoring the transportation infrastructure in the area is central to economic recovery. General engineering is required in order to initiate immediate improvement of the transportation network. These networks enable freedom of maneuver, logistics support, and the movement of personnel and material to support ongoing operations.

Civil Support Operations

4-32. The Armed Forces of the United States are authorized under certain conditions to provide assistance to U.S. civil authorities. This assistance is known as civil support within the defense community. Civil support is Department of Defense support to U.S. civil authorities for domestic emergencies and for designated law enforcement and other activities. This support usually entails Class III, subsistence, medical services, and general engineering support.

4-33. Civil support will always be in support of a lead federal agency such as the Department of Homeland Security. Requests for assistance from another agency may be predicated on mutual agreements between agencies or stem from a Presidential designation of a federal disaster area or a federal state of emergency. The military typically only responds after the resources of other federal agencies, state and local governments, and NGOs have been exhausted or when specialized military assets are required.

4-34. Within the Joint Staff J-3, the Joint Director of Military Support (JDOMS) serves as the Action Agent for Assistant Secretary of Defense for Homeland Defense (ASD (HD)) who has EA responsibilities for Defense Support of Civil Authorities (DSCA). JDOMS ensures the performance of all DSCA planning and execution responsibilities. JDOMS conducts planning and prepares warning and execution orders that task DOD resources in response to specific requests from civil authorities.

4-35. The National Response Framework (NRF) is an all-discipline, all-hazards plan that establishes a single, comprehensive framework for the management of domestic incidents. It provides the structure and mechanisms for the coordination of Federal support to State and local incident managers and for exercising direct Federal authorities and responsibilities. Through the NRF, FEMA assigns emergency support functions (ESF) to the appropriate federal agencies. The Army maintains a permanently assigned Defense Coordinating Officer in each FEMA region to plan, coordinate, and integrate DSCA with local, state, and federal agencies. DOD is the Primary Coordinating Agency for ESF 3 (Public Works and Engineering) and when requested, and upon approval of the SECDEF, DOD provides DSCA during domestic incidents and is considered a support agency to all ESFs. These include among others transportation, public health and medical services, and public safety and security.

4-36. The Army National Guard often acts as a first responder on behalf of state authorities when functioning under Title 32 U.S. Code authority or while serving on State active duty. National Guard sustainment systems may vary from state to state but are similar on a basic level. Generally, National Guard armories stock much of the logistics material required for initial 72-hour operations. Fuel and other expendable classes of supply are replenished using a combination of local, commercial sources, and State government sources. During catastrophic disaster response, State National Guard units eventually require Federal sustainment augmentation.

PREPARING FOR SUSTAINMENT OF OPERATIONS

4-37. Preparation for the sustainment of operations consists of activities performed by units to improve their ability to execute an operation. Preparation includes but is not limited to plan refinement, rehearsals, intelligence, surveillance and reconnaissance, coordination, inspections, and movements (FM 3-0). For sustainment to be effective, several actions and activities are performed across the levels of war to properly prepare forces for operations.

NEGOTIATIONS AND AGREEMENTS

4-38. Negotiating HNS and theater support contracting agreements may include pre-positioning of supplies and equipment, civilian support contracts, OCONUS training programs, and humanitarian and civil assistance programs. These agreements are designed to enhance the development and cooperative solidarity of the host nation and provide infrastructure compensation should deployment of forces to the target country be required. The pre-arrangement of these agreements reduces planning times in relation to contingency plans and operations.

4-39. Negotiation of agreements enables access to HNS resources identified in the requirements determination phase of planning. This negotiation process may facilitate force tailoring by identifying available resources (such as infrastructure, transportation, warehousing, and other requirements) which if not available would require deploying additional sustainment assets to support.

OPERATIONAL CONTRACT SUPPORT

4-40. Operational contract support plays an ever increasing role in the sustainment of operations and is an integral part of the overall process of obtaining support. Contract support is used to augment other support capabilities by providing an additional source for required supplies and services. Because of the importance and unique challenges of operational contract support, commanders and staffs need to fully understand their role in managing contract support in the AO.

4-41. The requiring activity (normally brigade through ASCC level units), in close coordination with the supporting contracting unit/office or Team LOGCAP-Forward, must be able to describe what is needed to fulfill the minimum acceptable standard for the government. A detailed description of the requirement is instrumental in allowing the contracting officer to create a solicitation against which bidders can submit a proposal and successfully deliver in accordance with the terms of the contract. As part of this process, the requiring activity is responsible to do basic research to ensure that the required support is not available through the non-commercial means (organic military support, multinational, and/or HNS sources) and to determine if there are any commercial standards for the item of support along with potential local sources of support.

4-42. The requiring activity, with support and guidance from the supporting contracting office and Team LOGCAP-Forward staff, is also required to develop an independent government estimate and, for service contracts, a performance work statement (sometimes referred to as a statement of work [SOW]). The requiring activity then must obtain staff and command approval of the requirements packet. While the supporting contracting unit will provide requirements development process advice and guidance, it is ultimately the requiring activity's responsibility to ensure that this packet is "acquisition ready" developed, staffed, approved, prioritized, and funded in accordance with local command policies.

4-43. Once the contract support request is approved, the requiring activity must also be prepared to provide government oversight assistance to the contracting organization, that includes at a minimum, qualified contracting officer representative support for all service contract and receiving official support (for community contracts). Finally, all requiring activities must ensure contract support is properly integrated into their operations in the areas of government furnished support, contractor accountability, force protection measures, and so forth (JP 4-10, FMI 4-93.41 and FMI 4-93.42).

4-44. An important capability for the commander is to incorporate contract support with operational reach. The major challenge is ensuring that theater support and external support contracts are integrated with the overall sustainment plan. It is imperative that the TSC/ESC SPO and the ASCC G-4 coordinate with the supporting CSB. The CSB assists the Theater Army G-4 to develop the contracting support integration plans. The CSB commands contracting deployed units to support those plans. Close coordination with the TSC and CSB is necessary to ensure proper execution of the contract support integration plan.

Theater Support Contracts

4-45. Theater support contracts assist deployed operational forces under prearranged contracts or contracts awarded in the AO by contracting officers under the C2 of the CSB. Theater-support contractors acquire goods, services, and minor construction support, usually from the local commercial sources, to meet the immediate needs of operational commanders. Theater support contracts are typically associated with contingency contracting. When this support involves a service contract, the unit must be prepared to provide a contracting officer representative.

External Support Contracts

4-46. External support contracts provide a variety of support functionalities to deployed forces. External support contracts may be prearranged contracts or contracts awarded during the contingency itself to support the mission and may include a mix of U.S. citizens, third-country nationals, and local national subcontractor employees. The largest and most commonly used external support contract is LOGCAP. This Army program is commonly used to provide life support, transportation support, and other support functions to deployed Army forces and other elements of the joint force as well. Depending on METT-TC factors, the TSC will often serve as the requiring activity for mission related LOGCAP support requirements. If designated by the Army forces as the priority unit for LOGCAP support, the TSC would normally be augmented by an USAMC logistics support officer from Team LOGCAP-Forward.

System Support Contracts

4-47. System support contracts are pre-arranged contracts by the USAMC LCMCs and separate ASA(ALT) program executive and product/project management offices. The AFSB coordinates the administration and execution of system contracts within an AO in coordination with LCMC and separate ASA(ALT) program executive and product/project management offices. Supported systems include, but are not limited to, newly fielded weapon systems, C2 infrastructure (such as the Army Battle Command Systems (ABCS) and standard Army management information systems (STAMIS)), and communications equipment. System contractors, made up mostly of U.S. citizens, provide support in garrison and may deploy with the force to both training and real-world operations. They may provide either temporary support during the initial fielding of a system, called interim contracted support, or long-term support for selected materiel systems, often referred to as contractor logistics support.

Army Pre-positioned Stocks (APS)

4-48. The APS program is a key Army strategic program. APS is essential in facilitating strategic and operational reach. USAMC executes the APS program and provides accountability, storage, maintenance, and transfer (issue and receipt) of all equipment and stocks (except medical supplies and subsistence items) (FM 1-01). Medical APS stocks are managed by U.S. Army Medical Materiel Agency for the Office of the

Surgeon General and subsistence items are managed for the Army by DLA. The reserve stocks are intended to provide support essential to sustain operations until resupply lines of communication can be established. Prepositioning of stocks in potential theaters provides the capability to rapidly resupply forces until air and sea lines of communication are established. Army prepositioned stocks are located at or near the point of planned use or at other designated locations. This reduces the initial amount of strategic lift required for power projection, to sustain the war fight until the LOC with CONUS is established, and industrial base surge capacity is achieved (FM 3-35.1).

4-49. The four categories of APS are: prepositioned unit sets, Operational Projects stocks, Army War Reserve Sustainment stocks, and War Reserve Stocks for Allies (FM 3-35.1), as discussed below.

Prepositioned Unit Sets

4-50. Prepositioned Unit Sets consist of prepositioned organizational equipment (end items, supplies, and secondary items) stored in unit configurations to reduce force deployment response time. Materiel is prepositioned ashore and afloat to meet the Army's global prepositioning strategic requirements of more than one contingency in more than one theater of operations.

Operational Projects Stocks

4-51. Operational projects stocks are materiel above normal table of organization and equipment (TOE), table of distribution and allowances (TDA), and common table of allowance (CTA) authorizations, tailored to key strategic capabilities essential to the Army's ability to execute force projection. They authorize supplies and equipment above normal modified TOE authorizations to support one or more Army operation, plan, or contingency. They are primarily positioned in CONUS, with tailored portions or packages prepositioned overseas and afloat. The operational projects stocks include aerial delivery, MA, and Force Provider (FP) base camp modules.

Army War Reserve Sustainment Stocks

4-52. Army War reserve sustainment stocks are acquired in peacetime to meet increased wartime requirements. They consist of major and secondary materiel aligned and designated to satisfy wartime sustainment requirements. The major items replace battle losses and the secondary items provide minimum essential supply support to contingency operations. Stocks are prepositioned in or near a theater of operations to reduce dependence on strategic lift in the initial stages of a contingency. They are intended to last until resupply at wartime rates or emergency rates are established.

War Reserve Stocks for Allies

4-53. War Reserve Stocks for Allies (WRSA) is an Office of the Secretary of Defense (OSD)-directed program that ensures U.S. preparedness to assist designated allies in case of war. The United States owns and finances WRSA assets and prepositions them in the appropriate theater. The United States positions APSs per Figure 4-3.

APS-1 (CONUS)	**Operational Project (OPROJ) stocks and Army War Reserve Sustainment (AWRS) stocks.**
APS-2 (Europe)	**Contingency unit sets, OPROJ, and WRSA.**
APS-3 (Afloat)	**Unit sets, OPROJ, and AWRS.**
APS-4 (Pacific and Northeast Asia)	**Unit sets, OPROJ, AWRS, and WRSA.**
APS-5 (Southwest Asia)	**Unit sets, OPROJ, and AWRS.**

Figure 4-3. Army Pre-positioned Stocks (APS) Locations

4-54. Land-based APS in Korea, Europe, or Southwest Asia allow the early deployment of a BCT to those locations. These prepositioned sets of equipment are essential to the timely support of the U.S. National military strategy in the areas of U.S. national interest and treaty obligations. Fixed land-based sites store Army prepositioned sets of BCT equipment, operational projects stocks, and sustainment stocks. Land-based sets can support a theater lodgment to allow the off-loading of Army pre-positioned afloat equipment and can be shipped to support any other theater worldwide (see FM 3-35.1).

4-55. The Automated Battlebook System (ABS) contains details on each APS program. G-3 planners and unit movement officers use ABS to identify equipment in the categories to accompany troops and not authorized for prepositioning. ABS also provides a consolidated list of all APS stockpile inventories. ABS supports deployment planning by providing the deploying unit with a contingency-updated database for all APS equipment and selected supplies in prepositioned locations. Army Field Support Brigades coordinate APS support to include command over AFSBs responsible to maintain and issue and account for APS unit equipment and supplies.

4-56. Army prepositioned afloat (APA) is the expanded reserve of equipment for an infantry brigade combat team (IBCT), theater-opening sustainment units, port-opening capabilities, and sustainment stocks aboard forward-deployed prepositioned afloat ships. APA operations are predicated on the concept of airlifting an Army IBCT with sustainment elements into a theater to link up with its equipment and supplies prepositioned aboard APA ships (see FM 3-35.1).

THEATER OPENING

4-57. Theater opening (TO) is the ability to rapidly establish and initially operate ports of debarkation (air, sea, and rail), to establish the distribution system and sustainment bases, and to facilitate port throughput for the reception, staging, and onward movement of forces within a theater of operations. Preparing for efficient and effective TO operations requires unity of effort among the various commands and a seamless strategic-to-tactical interface. It is a complex joint process involving the GCC and strategic and joint partners such as USTRANSCOM, its components, and DLA. Working together, TO functions set the conditions for effective support and lay the groundwork for subsequent expansion of the theater distribution system. It comprises many of the sustainment functions including, human resources (including Military Mail Terminal Team), FM, HSS, engineering, movement (air/land/water transport, inland terminal operations), materiel management, maintenance, and contracting.

4-58. When given the mission to conduct TO, a Sust Bde is designated a Sust Bde (TO) and a mix of functional battalions and multi-functional CSSBs are assigned based on mission requirements. The Sust Bde HQ staff may be augmented with a Transportation Theater Opening Element to assist in managing the TO mission. The augmentation element provides the Sust Bde with additional manpower and expertise to C2 TO functions, to conduct transportation planning, and provide additional staff management capability for oversight of reception, staging, onward movement, and integration (RSOI) operations, port operations, node and mode management, intermodal operations, and movement control. The Sust Bde will participate in assessing and acquiring available host nation (HN) infrastructure capabilities and contracted support (see FM 4-93.2).

JOINT DEPLOYMENT DISTRIBUTION OPERATIONS CENTER (JDDOC)

4-59. Also critical to the TO effort is the JDDOC. The JDDOC mission is to improve in-transit visibility and to support the geographic CCDR's operational objectives. The operational objective is accomplished by synchronizing and optimizing the interface of intertheater and intratheater distribution to integrate the proper flow of forces, equipment, and supplies. The JDDOC, under the control and direction of the GCC, plans and coordinates deployment and redeployment and strategic distribution operations. The JDDOC is an integral component of the GCC staff, normally under the direction of the GCC Director of Logistics (J4). However, GCC's can place the JDDOC at any location required or under the OPCON of another entity in the GCC area of responsibility. The JDDOC will coordinate with the TSC/ESC. However, on small scale operations, the JDDOC may coordinate directly with a Sust Bde operating as the senior Army LOG C2 HQ in the theater of operations. The JDDOC is directly linked to USTRANSCOM and provides strategic visibility.

PORT OPENING

4-60. Port opening and port operations are critical components for preparing TO. Commanders and staffs coordinate with the HN to ensure sea ports and aerial ports possess sufficient capabilities to support arriving vessels and aircraft. USTRANSCOM is the port manager for deploying U.S. forces (see FM 55-50 and 55-60).

Joint Task Force Port Opening (JTF-PO)

4-61. The JTF-PO is a joint capability designed to rapidly deploy and initially operate aerial and sea ports of debarkation, establish a distribution node, and facilitate port throughput within a theater of operations. The JTF-PO is a standing task force that is a jointly trained, ready set of forces constituted as a joint task force at the time of need.

4-62. The JTF-PO facilitates joint RSO&I and theater distribution by providing an effective interface with the theater JDDOC and the Sust Bde for initial aerial port of debarkation (APOD) operations. The JTF-PO is designed to deploy and operate for up to 60 days. As follow-on theater logistic capabilities arrive, the JTF-PO will begin the process of transferring mission responsibilities to arriving Sust Bde forces or contracted capabilities to ensure the seamless continuation of airfield and distribution operations.

SEAPORTS

4-63. The SDDC is the single port manager (SPM) for all common user seaports of debarkation (SPOD). The SPM may have OPCON of a port support activity (PSA) provided by any unit. The PSA assists in moving unit equipment from the piers to the staging/marshaling/loading areas, assisting the aviation support element with movement of helicopters in preparation for flight from the port, providing limited maintenance support for equipment being offloaded from vessels, limited medical support, logistics support, and security for port operations.

4-64. Ideally, the SPOD will include berths capable of discharging Large Medium Speed Roll-on/Roll-off ships. The SPOD can be a fixed facility capable of discharging a variety of vessels, an austere port requiring ships to be equipped with the capability to conduct their own offloading, or beaches requiring the conducting of

Logistics Over the Shore (LOTS) operations. Whatever the type of SPOD, it should be capable of accommodating a HBCT.

4-65. When vessels arrive at the SPOD, the port manager is responsible for discharging the unit equipment, staging the equipment, maintaining control and in-transit visibility (ITV), and releasing it to the unit. The port commander remains responsible for unit equipment and supplies until they reach the staging area where arriving units assume responsibility for their supplies and equipment. This includes minimum standards that are critical for the physical security/processing of DOD sensitive conventional arms, ammunition, and explosives, including non-nuclear missiles and rockets.

4-66. The Movement Control Team (MCT) manages the processing of the units' equipment for onward movement. Their actions are based on advanced manifests received via the Worldwide Port System, available transportation, theater priorities, tactical situation, and throughput capacity.

4-67. The Theater Gateway Personnel Accounting Team and supporting HR company and platoons will normally operate at the SPOD/APOD. The MCT that has responsibility for the SPOD, coordinates personnel accounting with the supporting CSSB or Sust Bde for executing life support functions (billeting, feeding, transportation, and so forth) for personnel who are transiting into or out of the theater.

AERIAL PORTS

4-68. The APOD is an airfield that has been designated for the sustained air movement of personnel and materiel. It is designated an APOD by the supported CCDR in coordination with USTRANSCOM. Reception at the APOD is coordinated by the senior logistics commander and executed by (an Air Force) Contingency Response Group/Element, a MCT, an arrival/departure control group (A/DACG), or both, depending on the magnitude of the operations. The MCT and/or A/DACG must be in the lead elements of the transported force. Augmentation with cargo transfer companies, cargo documentation teams, theater support contractor, and HNS is desired to rapidly clear the port. The port MCT has the mission of coordinating transport services for the APOD and ensuring quick clearance of cargo movements into and out of the APOD. Both Air Force and Army have responsibilities at an APOD.

4-69. Air Mobility Command (AMC) is the SPM for all common user APODs. Ideally, the APOD will provide runways of varying capacity, cargo handling equipment, adequate staging areas, multiple links to the road and rail network, and a qualified work force. The SPM has OPCON of an A/DACG provided by the Sust Bde that has the TO mission. The A/DACG assists in moving unit equipment from the aircraft to the staging/marshaling/loading areas. It also assists the aviation support element with movement of helicopters in preparation for flight from the APOD.

4-70. The A/DACG coordinates with the MCT for clearance of personnel and equipment from the APOD. The Air Force and the Inland Cargo Transfer Company ICTC do the offloading of aircraft and move them to the staging area for onward movement. The ICTC discharges, loads, and transships cargo at air, rail, or truck terminals.

WAREHOUSE AND BILLETING AND OTHER SUPPORT

4-71. Warehousing, billeting, and other infrastructure capabilities must be identified at each port of debarkation (POD) prior to the arrival of forces in theater. Any limitations influence the efficiency of the entire sustainment system. Host nation infrastructure such as electrical power grids, sanitation, bulk petroleum, oil, and lubricant (POL) availability, POL 'Tank Farms', and potable water sources and facilities are important to the successful employment and deployment of forces.

4-72. Force provider is one system for providing life support for transient forces deploying to operations. Force provider can be configured in a 600 person life support capability. It can be configured for transport in 150 person increments that provide environmentally controlled billeting, feeding, and field hygiene (laundry, shower, and latrine) capabilities. Add on capabilities include: cold weather kit; prime power connection kit; and morale, welfare, and recreation kit.

Medical Logistics Support

4-73. Medical units must be capable of operations immediately upon arrival and initial entry of forces. Therefore, medical logistics support must be included in planning for port opening and early entry operations. Medical logistics support to arriving forces includes Class VIII sustainment of primary medical care (sick call), including support to combat units so that organic medical supply levels are not depleted during RSO&I. Medical logistics also includes management of special medical materiel, such as medical chemical defense materiel, special vaccines, and other medical materiel under the control of the ASCC surgeon. Port operations may also include the issue of medical unit sets from APS and the integration of potency and dated pharmaceuticals, refrigerated, and controlled substances with those assemblages. In almost every operation, lessons learned reflect that theater MEDLOG units must also provide Class VIII materiel for unit shortages that were not filled prior to unit deployment (see FM 4-02.1).

Rehearsals and Training

4-74. Rehearsals are a vital component of preparing for operations. Large rehearsals require considerable resources, but provide the most planning, preparation, and training benefit. Depending on circumstances, units may conduct a reduced force or full dress rehearsal. The integration of sustainment and operational rehearsals are preparation activities. FM 5-0 describes the following rehearsals:

- Rock Drill. Rock drills allow key leaders to rehearse operational concepts prior to execution.
- Full Dress. Full rehearsals help Soldiers to clearly understand what is expected of them and gain confidence in their ability to accomplish the mission.
- Support Rehearsals. Support rehearsals complement preparations for the operation. They may be conducted separately and then combined into full dress rehearsals.
- Battle Drills and Standing Operating Procedures (SOP) Checks. A battle drill or SOP rehearsal ensures that all participants understand a technique or a specific set of procedures.

EXECUTING SUSTAINMENT OPERATIONS

4-75. Execution means putting a plan into action by applying combat power to accomplish the mission and using situational understanding to assess progress and make adjustments (FM 3-0). It focuses on concerted actions to seize, retain, and exploit the initiative. Execution of sustainment operations includes supporting force projection, basing, distribution, and reconstitution of forces. The provision of sustainment maintains combat power and prolongs endurance.

Strategic and Operational Reach and Endurance

4-76. Strategic reach is the distance a Nation can project decisive military power against complex, adaptive threats operating anywhere. Operational reach is the distance and duration across which a unit can successfully employ military capabilities. The ability to conduct strategic and operational reach combines joint military capabilities—air, land, maritime, space, special operations, and information systems with those of the other instruments of national power.

4-77. Sustainment enables strategic and operational reach. It provides joint forces with the lift, materiel, supplies, health support, and other support functions necessary to sustain operations for extended periods of time. Army forces require strategic sustainment capabilities and global distribution systems to deploy, maintain, and conduct operations anywhere with little or no advanced notice.

4-78. Army forces increase the joint force's strategic reach by securing and operating bases in the AOR. However, Army forces depend on joint-enabled force projection capabilities to deploy and sustain them across intercontinental distances. In many instances, land operations combine direct deployment with movements from intermediate staging bases located outside the operational area.

4-79. Extending operational reach is a paramount concern for commanders. To achieve the desired end state, forces must possess the necessary operational reach to establish and maintain conditions that define success.

Commanders and staffs increase operational reach through deliberate, focused operational design, and the appropriate sustainment to facilitate endurance.

4-80. Endurance stems from the ability to maintain, protect, and sustain forces, regardless of how far away they are deployed, how austere the environment, or how long land power is required. Endurance is enabled by an Army distribution system that provides forces with a continuous flow of sustainment. A discussion of the Army distribution system will occur later in this chapter.

FORCE PROJECTION

4-81. Force projection is the military element of national power that systemically and rapidly moves military forces in response to requirements across the spectrum of conflict. It includes the processes of mobilization, deployment, employment, sustainment, and redeployment of forces. These processes are a continuous, overlapping, and repeating sequence of events throughout an operation. Force projection operations are inherently joint and require detailed planning and synchronization.

4-82. Sustainment to force projection operations is a complex process involving the GCC, strategic and joint partners such as USTRANSCOM, and transportation component commands like AMC, MSC, SDDC, USAMC, DLA, Service Component Commands, and Army generating forces.

Mobilization

4-83. Mobilization is the process of bringing the armed forces to a state of readiness in response to a contingency. Sustainment builds and maintains force readiness. Army generating sustainment forces in the execution of its Title 10 mission prepare Army forces for full spectrum operations. Upon alert for deployment generating force sustainment organizations, ensure Army forces are manned, equipped, and meet all Soldier readiness criteria. IMCOM operates Army installations that serve as deployment platforms. Army active and reserve component units mobilize from Army installations that ensure Soldiers, equipment, materiel, and health Soldier readiness are verified prior to deployment.

Deployment

4-84. Deployment is the movement of forces to an operational area in response to an order. Sustainment is crucial to the deployment of forces. It requires joint and Service sustainment capabilities to strategically move and maintain deploying forces. Joint transportation assets including air and sealift capabilities provide the movement capabilities for the Army. Army forces are moved to APOEs and SPOEs generally by commercial means to begin the deployment process. Sustainment commands monitor and track unit deployment status and provide information to Army HQ.

Employment

4-85. Employment is the conduct of operations to support a JFC. It prescribes how to apply force and/or forces to attain specified national strategic objectives. Employment encompasses a wide array of operations—including but not limited to—entry operations, decisive operations, and post-conflict operations. Employment includes RSO&I, through sustainment preparation activities discussed earlier, Army forces which are able to enter established ports, assemble units for operations, and move personnel and equipment to operational areas.

Sustainment

4-86. Sustainment is the provisioning of logistics, personnel services, and health services support to maintain forces until mission completion. It is through sustainment that Army forces are able to accomplish and complete assigned missions.

Redeployment

4-87. Redeployment is the return of forces and materiel to the home or mobilization station or to another theater. Redeployment is just as important as deployment. In many instances, it is more difficult. It requires retrograde of logistics, personnel services, and health services support. For example, redeploying units must undergo U.S. Customs and Agriculture inspections prior to departure. Reuniting unit personnel and their equipment at their home station triggers the start of the lifecycle management process (reset/train, ready, and available to deploy).

BASING

4-88. A base is a locality from which operations are projected or supported (JP 1-02). The base includes installations and facilities that provide sustainment. Bases may be joint or single Service areas. Commanders often designate a specific area as a base and assign responsibility for protection and terrain management with the base to a single commander. Units located within the base are under the tactical control of the base commander, primarily for the purpose of facilitating local base defense. Within large bases, controlling commanders may designate base clusters for mutual protection and C2 (see Figure 4-4).

INTERMEDIATE STAGING BASES (ISB)

4-89. An ISB is a secure base established near, but not in, the AOR through which forces and equipment deploy (FM 3-0). While not a requirement in all situations, the ISB may provide a secure, high-throughput facility when circumstances warrant. The commander may use an ISB as a temporary staging area en route to a joint operation, as a long-term secure forward support base, and/or secure staging areas for redeploying units, and noncombatant evacuation operations (NEO).

4-90. An ISB is task organized to perform staging, support, and distribution functions as specified or implied by the CCDR and the Theater Army operations order. The ISB task organization is dependent on the operational situation and the factors of METT–TC. It may provide life support to staging forces in transit to operations or serve as a support base supporting the theater distribution plan.

4-91. As a support base, an ISB may serve as a transportation node that allows the switch from strategic to intratheater modes of transportation. Whenever possible an ISB takes advantage of existing capabilities, serving as a transfer point from commercial carriers to a range of tactical intratheater transport means that may serve smaller, more austere ports. Army forces may use an ISB in conjunction with other joint force elements to pre-position selected sustainment capabilities. ISB personnel may perform limited sustainment functions, such as materiel management and selected sustainment maintenance functions.

FORWARD OPERATING BASES

4-92. Forward operating bases extend and maintain the operational reach by providing secure locations from which to conduct and sustain operations. They not only enable extending operations in time and space; they also contribute to the overall endurance of the force. Forward operating bases allow forward deployed forces to reduce operational risk, maintain momentum, and avoid culmination.

4-93. Forward operating bases are generally located adjacent to a distribution hub. This facilitates movement into and out of the operational area while providing a secure location through which to distribute personnel, equipment, and supplies.

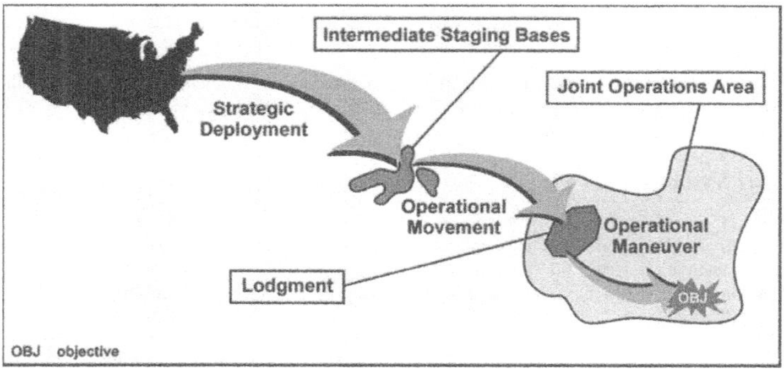

FIGURE 4-4 BASING

DISTRIBUTION

4-94. Distribution is the key component for executing sustainment. It is based on a distribution system defined as that complex of facilities, installations, methods, and procedures designed to receive, store, maintain, distribute, and control the flow of military materiel between point of receipt into the military system and point of issue to using activities and units.

4-95. The Joint segment of the distribution system is referred to as global distribution. It is defined as the process that synchronizes and integrates the fulfillment of joint requirements with the employment of joint forces. It provides national resources (personnel and materiel) to support the execution of joint operations. The ultimate objective of the process is the effective and efficient accomplishment of joint operations. The Army segment of the distribution system is theater distribution. Theater distribution is the flow of equipment, personnel, and materiel within theater to meet the CCDR's mission. The theater segment extends from the ports of debarkation or source of supply (in theater) to the points of need (Soldier) (FM 4-01.4).

4-96. Theater distribution is enabled by a distribution management system. Distribution management **is the function of synchronizing and coordinating a complex of networks (physical, communications, information, and resources) and the sustainment WFF (logistics, personnel services, and HSS) to achieve responsive support to operational requirements**. Distribution management includes the management of transportation and movement control, warehousing, inventory control, materiel handling, order administration, site and location analysis, packaging, data processing, accountability for people and equipment, and communications. It involves activities related to the movement of materiel and personnel from source to end user, as well as retrograde operations.

DISTRIBUTION MANAGEMENT CENTER (DMC)

4-97. Theater distribution management is conducted by the DMCs located within the support operations (SPO) section of the TSC and ESC. The DMC develops the theater distribution plan and monitors distribution performance in coordination with strategic distribution process owners and the support operations staffs in Sust Bdes and BSBs, This coordination ensures timely movement and retrograde of sustainment within the CCDR's area of responsibility. The DMC coordinates distribution with the HRSC and ASCC G-4/G-1/G-8 to ensure personnel and resources are linked. It exercises staff supervision of movement control units in a theater.

4-98. The DMC orchestrates the distribution of all classes of supply and manages all aspects of theater distribution by maintaining visibility of requirements, managing the capacity of the system, and controlling the execution of distribution operations. The DMC considers the impact of unit movement requirements on the distribution system. It provides current information on location of mode assets and movement of critical supplies along main supply routes. They provide staff recommendations to direct, redirect, retrograde, and cross-level resources to meet the distribution mission and user mission requirements.

4-99. The distribution management of medical materiel is accomplished by a support team from the MLMC. The MLMC support team collocates with the DMC of the TSC/ESC to provide the MEDCOM (DS) with visibility and control of all Class VIII.

IN-TRANSIT VISIBILITY (ITV)

4-100. ITV is the ability to track the identity, status, and location of DOD units, and non-unit cargo (excluding bulk petroleum, oils, and lubricants) and passengers; patients and personal property from origin to consignee, or destination across the range of military operations (JP 3-35). This includes force tracking and visibility of convoys, containers/pallets, transportation assets, other cargo, and distribution resources within the activities of a distribution node.

4-101. Visibility begins at the point where personnel and materiel enter the distribution system. Data concerning personnel and materiel are entered into the appropriate automated system. This data is updated by subsequent ITV systems until it reaches its final destination (automation systems are listed in Appendix A). The information is accessible to all users regardless of the military service or echelon of command. Figure 4-5 shows the ITV process.

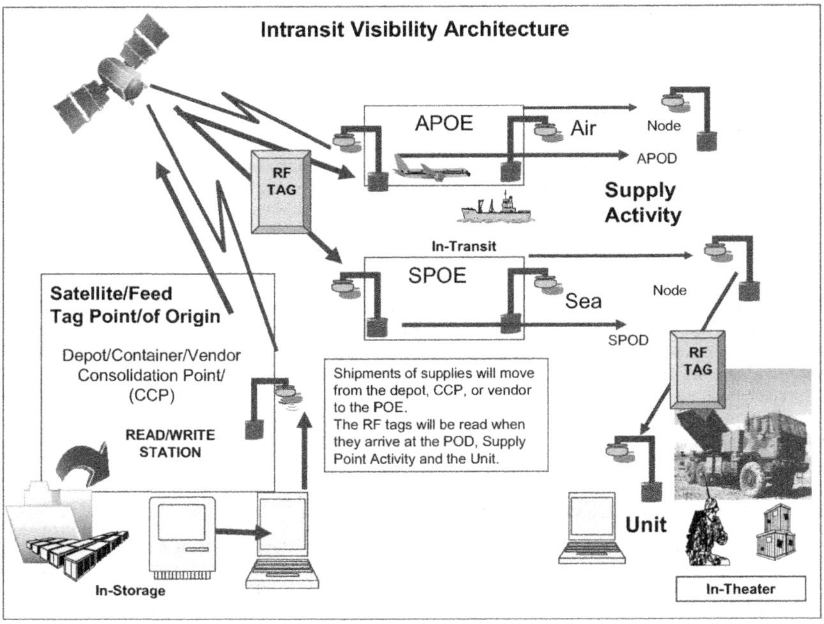

Figure 4-5. ITV Architecture

4-102. ITV provides the distribution manager the ability to assess how well the distribution process is responding to supported force needs. Distribution managers gain and maintain visibility (items, personnel, units, transition hubs, and transport modes) at the earliest practical point in the management process. This allows managers to operate with timely information. ITV of personnel and materiel is continuous throughout the distribution process.

Retrograde of Materiel

4-103. Retrograde of materiel is the return of materiel from the owning/using unit back through the distribution system to the source of supply, directed ship-to location and/or point of disposal (FM 4-01.4). Retrograde includes turn-in/classification, preparation, packing, transporting, and shipping. To ensure these functions are properly executed, commanders must enforce supply accountability and discipline and utilize the proper packing materials. Retrograde of materiel can take place as part of theater distribution operations and as part of redeployment operations. Retrograde of materiel must be continuous and not be allowed to build up at supply points/nodes.

4-104. Early retrograde planning is essential and necessary to preclude the loss of materiel assets, minimize environmental impact, and maximize use of transportation capabilities. Planners must consider environmental issues when retrograding hazardous materiel.

4-105. Contractor or HNS may be used in the retrograde of materiel. This support is planned and negotiated early in the operation. HNS must be identified early enough to ensure they are properly screened and present no security risk. Leaders at all levels are responsible for the adherence of all policies and safety measures by contractors and HNS.

4-106. The theater distribution system provides the ASCC the ability to monitor and manage retrograde materiel through the system. Retrograde materiel flows through the distribution system in the reverse order from the tactical to strategic levels. Retrograde materiel is consolidated at the lowest supply support activity (SSA) and reported up through the support operations for distribution instructions. When released by the maneuver commander, USAMC assumes responsibility for providing disposition instructions, accounting, and shipment of retrograde materiel from the theater.

4-107. An approved military customs inspection program must be in place prior to redeployment to pre-clear not only redeployment materiel but also the shipment of battle damaged equipment out of theater. The ASCC is responsible for establishing the customs inspection program to perform U.S. customs pre-clearance and United States Department of Agriculture inspection and wash down on all materiel retrograded to the United States in accordance with DOD 4500.9-R.

IN-THEATER RECONSTITUTION

4-108. **In-theater reconstitution is extraordinary actions that commanders take to restore a degraded unit to combat effectiveness commensurate with mission requirements and available resources**. In-theater reconstitution should be considered when the operational tempo, mission, or time, does not allow for replacements by an available unit. Reconstitution requires both generating and operating force involvement. Generally it should be conducted in a relatively low stress environment.

4-109. The combat readiness of the unit, mission requirements, risk, and the availability of a replacement unit are the keys for considering reconstitution operations. Commanders must closely evaluate the combat worthiness of a unit to determine whether a reconstitution operation should be ordered. He/She must also decide what type of reconstitution effort would be best for the organization based on METT-TC factors. The three major elements of in theater reconstitution are reorganization, regeneration, and rehabilitation.

Reorganization

4-110. **Reorganization is action to shift resources within a degraded unit to increase its combat effectiveness**. Commanders of all types of units at each echelon may conduct reorganization. Reorganization may be conducted when the operational tempo is such that the risk for removing a unit from the operation may jeopardize the mission. Depending on METT-TC factors there are two types of reorganization operations, immediate and deliberate.

Immediate Reorganization

4-111. Immediate reorganization is the quick and usually temporary restoring of degraded units to minimum levels of effectiveness. Normally the commander implements immediate reorganization in the combat position or as close to that site as possible to meet near term needs. Immediate reorganization consists of cross leveling personnel and equipment, matching weapon systems to crews, or forming composite units (joining two or more attrited units to form a single mission-capable unit).

Deliberate Reorganization

4-112. Deliberate reorganization is conducted when somewhat more time and resources are available. It usually occurs farther away from hostile activity than immediate reorganization. Procedures are similar to those for immediate reorganization. However, some replacement resources may be available. Also, equipment repair is more intensive and more extensive cross-leveling is possible.

REGENERATION

4-113. **Regeneration is the rebuilding of a unit. It requires large-scale replacement of personnel, equipment, and supplies.** These replacements may then require further reorganization. Regeneration involves reestablishing or replacing the chain of command and conducting mission essential training to get the regenerated unit to standard. Because of the intensive nature of regeneration, it occurs at a designated regeneration site after the unit disengages from operations. The regeneration site is normally situated in a relatively secure location.

4-114. Regeneration requires help from higher echelons and may include elements from the generating force, contract support, and HNS. Since regeneration typically requires large quantities of personnel and equipment, commanders carefully balance these needs against others in the command as well as the mission.

REHABILITATION

4-115. Rehabilitation is the processing, usually in a relatively quiet area, of units or individuals recently withdrawn from combat or arduous duty, during which units recondition equipment and are rested, furnished special facilities, filled up with replacements, issued replacement supplies and equipment, given training, and generally made ready for employment in future operations (JP 1-02).

4-116. Rehabilitation will most likely require similar high levels of support to execute. The main difference is that rehabilitation may occur when time is not a critical factor. If the conditions are such that forces will rotate in and out of operations on a regular basis, commanders may choose to establish a semi-permanent rehabilitation site.

SUSTAINMENT OF IN-THEATER RECONSTITUTION OPERATIONS

4-117. Regardless of the type of reconstitution operation, sustainment of the operation will be intense. Reconstitution of a unit will require involvement by most, if not all, of the sustainment functions to execute. Logistics support will require a full range of capabilities including field services, maintenance, supply, transportation, contract support, and general engineering. From a personnel service perspective, the full range of HRS will be required in addition to FM, legal, and religious support. Likewise, robust medical support will be required and in some circumstances combat and operational stress control may be a key element.

4-118. The Sust Bde will be crucial to providing the logistical support to reconstitution. It may require support from the AFSB, CSB, and USAMC capabilities to provide the full range of required sustainment. The MEDCOM (DS) will be the sustainment organization responsible for AHS support.

SUSTAINMENT OF INTERNMENT RESETTLEMENT OPERATIONS

4-119. The Army is DOD's EA for all detainee operations. Additionally, the Army is DOD's EA for long-term confinement of U.S. military prisoners. I/R operations are defined as operations that take or keep selected individuals in custody or control as a result of military operations to control their movement, restrict their activity, provide safety, and/or gain intelligence (FM 3-19.40).

4-120. I/R operations comprise those measures necessary to guard, protect, sustain, and account for people that are captured, detained, confined, or evacuated from their homes by the U.S. armed forces. I/R operations require detailed advanced planning to prevent the degradation of operational momentum while providing a safe and secure environment for prisoners. U.S. policy mandates that all individuals captured, interned, evacuated, or held by U.S. armed forces are treated humanely. This policy applies from the moment they are under the control of U.S. armed forces until they are released, repatriated, or resettled.

4-121. The Army MPs are tasked with coordinating shelter, protection, accountability, and sustainment for detainees. I/R functions, though under the sustainment WFF, address MP roles when managing detainees and dislocated civilians during operations.

4-122. Sustainment to I/R operations involves a wide range of support including logistics, personnel services, and medical treatment to detained persons. It encompasses providing all classes of supplies and materiel, health and personnel services, and general engineering support. General engineering provides horizontal and vertical construction, as well as repair and maintenance of the infrastructure (see FM 3-34.400).

- Personnel services include the FM of negotiable instruments of applicable detainees. They also coordinate for payroll, disbursement, and repatriation settlement processing (see FM 1-06).
- The Judge Advocates monitor the treatment of detainees ensuring that U.S. Soldiers are adhering to the applicable standards for detainee treatment.
- The chaplain or unit ministry team assists the commander in providing religious support for I/R operations. The chaplain or UMT: serves as the chaplain for detention facility personnel, which does not include detainees, advises the commander on detainee religious issues and support, serves as a moral and ethical advisor to the Detention Facility Commander, exercises supervision and control over Retained Personnel religious leaders within the facility, and is prohibited from privileged communications with detainees."
- AHS support ensures personnel in U.S. custody receive medical care consistent with the standard of medical care that applies to U.S. armed forces in the same area. Medical personnel are responsible for the health care of the interred and resettled population. Preventative medicine personnel inspect water for potability, conduct pest management activities, ensure food service facilities meet sanitation requirements, inspect facilities for sanitation, and monitor communicable diseases.

ASSESSING THE SUSTAINMENT OF OPERATIONS

4-123. Assessment is the continuous process that occurs throughout the operations process. Sustainment commanders and staffs monitor and evaluate the current situation and the progress of the operation and compare it with the concept of support, mission, and commander's intent. Based on their assessment, commanders direct adjustments to sustainment operations, ensuring that they remain focused on the mission and commander's intent.

4-124. The primary tools for assessing are the staff running estimates (see FM 6-0). A running estimate is a staff section's continuous assessment of current and future operations to determine if the current operation is proceeding according to the commander's intent and if future operations are supportable (see FM 3-0, FM 5-0, and FMI 5-0.1).

4-125. During planning, assessment focuses on understanding the current condition in the operational environment and developing relevant courses of action. During preparation and execution, it emphasizes evaluating progress toward the desired end state determining variance from expectation and determining the significance of these variances.

SUMMARY

4-126. Integrating the many elements of sustainment into operations is essential for mission success. The operations process of planning, preparation, execution, and assessment applies to sustainment in support of full spectrum operations.

4-127. Sustainment commanders integrate and synchronize sustainment with the operations plan. Sustainment commanders across all levels of war prepare to effectively execute the plan. Commanders must follow procedures to prepare for the execution of the operation. One of the means for preparation is sustainment preparation of the operational environment. This may entail host nation agreements, cross-service arrangements, and contracting. Other preparations include prepositioned stocks, facilities, ports, medical preparations, and an array of rehearsals. The execution of sustainment includes the deployment and distribution processes.

4-128. Integration of sustainment into operations optimizes operational reach and endurance. Distribution is the largest single process in the execution of sustainment operations and is controlled through distribution management centers and in-transit visibility enablers. The continual assessment of sustainment operations ensures mission success and allows commanders to adjust to changing situations as required.

Chapter 5
Functions of Sustainment

This chapter explains in greater detail, the functional elements of the sustainment WFF. The sustainment WFF is the related tasks and systems that provide support and services to ensure freedom of action, extend operational reach, and prolong endurance. This chapter explains the basic building blocks of sustainment, with its elements, fundamentals, and main activities. The broad functional areas were outlined and defined in Chapter 1.

SUPPLY AND FIELD SERVICES

5-1. Supply and Field Services are essential for enhancing Soldiers quality of life. Supply provides the materiel required to accomplish the mission. Field service provides life support functions, including field laundry, showers, light textile repair, FP, MA, aerial delivery support, food services, billeting, and sanitation.

FUNDAMENTALS OF SUPPLY OPERATIONS

5-2. Supply operations include the requisitioning, receipt, storage, issue, distribution, protection, maintenance, retrograde, and redistribution of supplies. Levels of supply are broadly classified under the levels of war as tactical, operational, and strategic.

- Tactical level supplies are those items provided to and carried within each maneuver or support brigade to sustain operational endurance. They also consist of those supplies held by Sust Bdes to provide area support.
- Operational supplies are theater stocks positioned to replenish tactical stocks, when strategic replenishment is not feasible.
- Strategic supplies are items under the control of strategic managers and are available for worldwide materiel release. These supplies are considered inventory in motion and part of the distribution system.

5-3. Supply operations with total asset visibility enablers merge the tactical, operational, and strategic levels into a seamless supply system. The automated management systems allow units to place their requests and assists sustainment units in providing responsive support in a timely manner. Table 5-1 lists the U.S classes of supply.

5-4. While munitions is a class of supply, it is unique due to the complexities of activities associated with its handling. Munitions are a dominant factor in determining the outcome of full spectrum operations. Munitions provide the means to defeat and destroy the enemy. Planning munitions support is considered and synchronized from strategic to tactical levels. The results of planning and integrating munitions operations is to ensure munitions arrive in the right quantities and proper types where and when needed.

5-5. The ASCC has overall responsibility for in-theater receipts, accountability, and management of munitions stocks. The ASCC is also responsible for establishing a Theater Support Area and Ammunition Supply Points (ASPs). It is also responsible for coordinating distribution between storage sites, forward Ammunition Transfer and Holding Point (ATHP), and direct issue to using units on an area support basis.

5-6. The ASP is run by an ordnance company assigned to a sustainment brigade. ASPs receive, store, issue, and maintain a one- to three-day supply of ammunition to meet a routine surge and emergency requirements for supported units. ASP stockage levels are based on tactical plans, availability of ammunition, and the threat to the supply operations.

5-7. ATHPs are the most mobile and responsive of all ASAs. Each BCT and selected support brigades are authorized an ATHP. It is located within the brigade support area (BSA) and is manned and operated by the ATHP section of the BSB distribution company. See FM 4-30.13 for more details of munitions support.

Class I	Subsistence, including health and welfare items.
Class II	Clothing, individual equipment, tentage, tool sets and tool kits, hand tools, administrative, and housekeeping supplies and equipment (including maps). This includes items of equipment, other than major items, prescribed in authorization/allowance tables and items of supply (not including repair parts).
Class III	POL, petroleum and solid fuels, including bulk and packaged fuels, lubricating oils and lubricants, petroleum specialty products; solid fuels, coal, and related products.
Class IV	Construction materials, to include installed equipment and all fortification/barrier materials.
Class V	Ammunition of all types (including chemical, radiological, and special weapons), bombs, explosives, mines, fuses, detonators, pyrotechnics, missiles, rockets, propellants, and other associated items.
Class VI	Personal demand items (nonmilitary sales items).
Class VII	Major items: A final combination of end products which is ready for its intended use: (principal item) for example, launchers, tanks, mobile machine shops, vehicles.
Class VIII	Medical material, including medical peculiar repair parts.
Class IX	Repair parts and components, including kits, assemblies and subassemblies, reparable and nonreparable, required for maintenance support of all equipment.
Class X	Material to support nonmilitary programs; such as, agricultural and economic development, not included in Class I through Class IX.

Table 5-1 Classes of Supply

FUNDAMENTALS OF FIELD SERVICES

5-8. Field services provide quality of life for Soldiers. Field services are made up of six functions discussed below.

Shower and Laundry

5-9. Shower and laundry capabilities resident within the Field Services Company are provided from the Sust Bdes for supported units as far forward as possible. The mission is to provide Soldiers a minimum of one weekly shower and up to 15 pounds of laundered clothing each week (comprising two uniform sets, undergarments, socks, and two towels). Shower and Laundry Clothing Repair Teams from the modular Quartermaster Field Services Company can be moved forward to provide field services for the BCT. The laundry and shower function does not include laundry decontamination support (see FM 3-11.5).

Food Preparation

5-10. Food preparation is a basic unit function and one of the most important factors in Soldier health, morale, and welfare. Virtually every type of unit in the force structure has some organic food service personnel.

5-11. The field feeding system assumes use of Meals Ready to Eat (MREs) for the first several days following deployment, followed by transition to prepared group feeding rations. The theater initially transitions from MREs to Unitized Group Rations. Then, as the operational situation permits, A-rations (fresh foods) are introduced into theater. This requires extensive sustainment expansion since it requires refrigeration, storage, distribution, and ice making. The standard is to provide Soldiers at all echelons three quality meals per day. Proper refuse and waste disposal is important to avoid unit signature trails and maintain field sanitation standards. See FM 4-20.2 for more details.

Water Production and Distribution

5-12. Water is an essential commodity. It is necessary for hydration, sanitation, food preparation, medical treatment, hygiene, construction, and decontamination. Support activities, such as helicopter maintenance, FP, and operation of medical facilities, consume large volumes of water. Classification of the water function is somewhat different from other commodities; it is both a field service and a supply function. Water purification is a field service. Quartermaster supply units normally perform purification in conjunction with storage and distribution of potable water which are supply functions. It is the users' responsibility to determine potable water requirements and submit them through supply channels.

5-13. Water supply units perform routine testing. However, water quality monitoring is primarily the responsibility of preventive medicine personnel of the MEDCOM (DS). The command surgeon ensures the performance of tests associated with water source approval, monitoring and interpreting test results. Each service provides its own water resource support. Typically, the Army, as directed by the JFC, provides support in a joint operation. AR 700-136 details the responsibilities of Army elements for water support.

5-14. Engineers play a major role in providing water to Army forces. They are responsible for finding subsurface water, drilling wells, and constructing, repairing, or maintaining water facilities. Geospatial engineers generate, manage, and analyze hydrologic data and work together with ground-survey teams and well-drilling teams to locate water sources.

5-15. The quantity of water required depends on the regional climate and the type and scope of operations. Temperate, tropic, and arctic environments normally have enough fresh surface and subsurface water sources to meet raw water requirements for the force. In arid regions, providing water takes on significantly greater dimensions. Soldiers must drink more water. Water requirements are significantly greater in areas, where demand is heavy for aircraft and vehicle washing, medical treatment, laundry and shower facilities, and where construction projects are conducted.

5-16. I/R operations may require a large amount of unanticipated bulk water consumption. Units must consider the potential absence of water capability in enemy units and the requirement for on-site sanitation, shower, delousing, and medical support for in-coming detainees. Since water is a critical commodity in arid regions, managers must strictly control its use. Commanders must establish priorities.

5-17. Because of the scarcity of potable water in some contingency areas, water support equipment may be prepositioned afloat. This allows initial support to a contingency force. Additional water equipment is available in CONUS depots to sustain operations. Most of this equipment is packaged for tactical transportability. Its configuration allows for throughput to the user with minimal handling in the AO.

Clothing and Light Textile Repair

5-18. Clean, serviceable clothing is essential for hygiene, discipline, and morale purposes. During peacetime, fixed facilities or field expedient methods normally provide clothing repair for short-duration exercises. During combat operations, they are provided as far forward as the brigade area.

5-19. Forces receive clothing support from a combination of units, HNS, and contractors. In low levels of hostilities, HNS and contractors may provide much of this support. LOGCAP offers considerable capability during the early deployment stages. A field service company provides direct support at the tactical level. The company has the modular capability of sending small teams as far forward as desired by the supported commander.

Aerial Delivery

5-20. Aerial delivery equipment and systems include parachute packing, air item maintenance, and rigging of supplies and equipment. This function supports both airborne insertions and airdrop/airland resupply. Airborne insertion involves the delivery of fighting forces, along with their supplies and equipment, to an objective area by parachute. Airdrop resupply operations apply to all Army forces. The airdrop function supports the movement of personnel, equipment, and supplies. It is a vital link in the distribution system and provides the capability of supplying the force even when land LOCs have been disrupted or terrain is too hostile, thus adding flexibility to the distribution system.

5-21. USAMC manages most airdrop equipment and systems (ADES). It maintains the national inventory control point and national maintenance point for ADES. At the operational level, the airdrop equipment repair and supply company provides supply and maintenance support to airdrop supply companies.

5-22. Aerial delivery support (ADS) companies provide airdrop resupply support in the corps/division area. They also provide personnel parachute support to units such as airborne and long range surveillance units. If the corps cannot support an airdrop request, it passes the request to a Sust Bde at theater level. Heavy Airdrop Supply Companies provide theater level support. Most of the supplies used for rigging by the ADS Company come directly from strategic level, bypassing the Airdrop Equipment Repair and Supply (AERS) Company. The AERS Company provides airdrop equipment supply and maintenance for the Heavy Airdrop Supply Company.

5-23. Airdrop resupply support must be flexible. Certain contingencies may require airdrop resupply from the beginning of hostilities. However, the requisite airdrop support structure is not likely to be in place due to other deployment priorities. In such cases, the operational commander should consider having a portion of the airdrop supply company deploy to the depot responsible for supply support to the contingency area. If forces require airdrop resupply before airdrop support units deploy to the theater, the units may rig supplies for airdrop at the depot. Supplies can then be flown directly to the airdrop location.

Mortuary Affairs (MA)

5-24. The MA program is a broadly based military program to provide for the necessary care and disposition of deceased personnel. Each service has the responsibility for returning remains and personal effects to CONUS. The Army is designated as the EA for the Joint MA Program. It maintains a Central Joint MA Office and provides general support to other services when requirements exceed their capabilities. The MA Program is divided into three subprograms:

- The Current Death Program provides mortuary supplies and associated services for permanently disposing remains and personal effects of persons for whom the Army is or becomes responsible.
- The Graves Registration Program provides search, recovery, initial identification, and temporary burial of deceased personnel in temporary burial sites. It also provides for the maintenance of burial sites and for the handling and disposing of personal effects.
- The Concurrent Return Program is a combination of the Current Death and Graves Registration Programs. This program provides for the search, recovery, and evacuation of remains to collection points and further evacuation to a mortuary. It provides for identification and preparation of remains at the mortuary and shipment to final destination as directed by next of kin.

5-25. MA units operate theater collection points, evacuation points, and personal effects depots. MA personnel initially process remains in theater. A MA Decontamination Collection Point may become operational whenever the threat of CBRN warfare exists. They then arrange to evacuate remains and personal effects, usually by air, to a CONUS POD mortuary. CONUS POD mortuaries provide a positive identification of the remains and prepare them for release in accordance with the desires of next of kin. MA processing points include mortuary affairs collection points (MACPs), theater mortuary evacuation points, mortuary affairs decontamination collection points, temporary interment sites, ID laboratories, and a port mortuary.

5-26. When directed by the CCDR, MA units establish cemeteries and provide for temporary interment of remains. MA units may also operate in-theater mortuaries, but they require personnel and equipment augmentation or HNS for embalming and other procedures (see JP 4-06 and FM 4-20.64).

5-27. A process and location for evacuation of personnel remains and equipment must be established. The responsibility for manning and running this activity must be done by the service responsible for the theater and have coordination and automation capability to process actions in the personnel automation systems as well as logistical systems. All personnel have clothing and other issue documents that must be cleared as well as personal effects that must be inventoried, cleaned, disposed of, and entered back into the system for issue. This is normally done by units, but also done at the Joint Personnel Effects Depot for those killed in action as well as wounded in action that have been evacuated and separated from their equipment.

TRANSPORTATION

5-28. Army transportation units play a key role in facilitating force projection and sustainment. Army transportation ensures that Army and joint forces that are projected globally are able to be sustained in operations. Transportation operations encompass the wide range of capabilities provided by transportation units and Soldiers. In joint operations, Army transportation units provide the full range of capabilities needed to allow joint and Army commanders to achieve operational objectives.

FUNDAMENTALS OF TRANSPORTATION MANAGEMENT

5-29. The Army transports personnel, cargo, and equipment by motor, rail, air, and water with organic or contract assets. While each situation may not be conducive to using a particular mode, the Army must be able to manage, operate, and supervise these modes of transport. Mode platforms include trucks, trains, containers, flatracks, watercraft, aircraft, and host nation assets. To successfully execute force projection operations and extend operational reach and endurance, Army transportation units must execute the following functions: movement control, terminal operations, and mode operations (see FM 55-1).

Movement Control

5-30. Movement control is the planning, routing, scheduling, controlling, coordination, and in-transit visibility of personnel, units, equipment, and supplies moving over line(s) of communication (LOC) and the commitment of allocated transportation assets according to command planning directives. It is a continuum that involves synchronizing and integrating sustainment efforts with other programs that span the spectrum of military operations. Movement control is a tool used to help allocate resources based on the CCDR's priorities and to balance requirements against capabilities (FM 4-01.30).

5-31. The five basic principles of movement control provide a basis for all transportation operations. These principles are:

- **Centralized Control and Decentralized Execution.** Centralized Control means that a focal point for transportation planning and resource allocation exists at each level of command involved in an operation. Decentralized execution of transportation missions means terminal and mode operators remain free to assign and control the specific transportation assets that will meet the requirement.
- **Regulated Movements**. Movement control authorities regulate moves to prevent terminal and route congestion and scheduling conflicts among Service components.
- **Fluid and Flexible Movements.** Transportation systems must provide the uninterrupted movement of personnel, supplies, and services. To do this, the system must be capable of rerouting and diverting traffic.
- **Effective Use of Carrying Capacity.** This principle is simple: keep transportation assets fully loaded and moving as much as the tactical situation permits.

- **Support Forward.** Support forward is throughput, which means rapid delivery of supplies and personnel as far forward as possible. It is dependent on fast, reliable transportation to move supplies and personnel as far forward as the tactical situation requires and permits.

Mode Operations

5-32. Mode operations and movement control elements working together match up the correct asset capability depending on cargo characteristics and required delivery time. Movement control sections coordinate transportation assets. When allocated, Army aviation assets for sustainment support direct coordination between the MCB and the Aviation Brigade is vital in providing responsive support. Requests for use of Air Force fixed winged aircraft for sustainment resupply requires coordination between the MCB and the theater airlift liaison officer. Airlift providers may be the Army, Navy, Air Force, MNFs, host nation military, or commercial aircraft.

Motor

5-33. Army motor transportation provides essential distribution capabilities for Army organizations. Army transportation units are the single largest provider of land surface movement within joint forces. Motor transportation includes organic, host nation, and contracted resources.

Rail

5-34. Rail is potentially the most efficient ground transportation method for hauling large tonnages. The Army has limited railway operating, construction, and repair capabilities. Rail capability may be provided through HNS. The Army augments HNS by providing personnel resources.

Air

5-35. Airlift is a mode of transportation. Wide-ranging sustainment needs within a theater require Air Force and Army airlift assets to support. Army utility helicopters provide support at the through movement control channels in response to mission requirements and the commander's priorities. Likewise, the U.S. Air Force provides intratheater airlift to all services within a theater through an allocation process on a routine basis or provides immediate support to operational requirements. While airlift is the preferred method of delivery, airdrop is a field service that can provide additional flexibility. It makes possible rapid resupply of critical items over extended distances directly to or near forward units.

Water

5-36. Army watercraft is a component of intratheater transportation. It can augment other modes when integrated with appropriate terminal operations or may be the primary means of transport in specific areas in a theater. Army watercraft move materiel and equipment over inland waterways, along theater coastlines, and within marine terminals. Their primary role is to support cargo discharge from strategic lift assets, conduct onward movement, and provide distribution of cargo and equipment from the SPOD to inland terminals and austere delivery points or retrograde materiel from those areas.

5-37. Watercraft can perform utility missions including patrolling, salvage, ship-to-shore transport of personnel, and harbormaster duties. Although not an Army watercraft mission, they can perform limited docking and undocking services for strategic transport vessels when required. The watercraft fleet consists of a variety of vessels such as landing craft, tug boats, floating cranes, barges, causeways, and associated equipment. Army watercraft are organized into companies and detachments which can operate under a variety of command relationships (such as attached to a Transportation Terminal Battalion or SDDC units (see FM 55-80).

Terminal Operations

5-38. Terminal operations are key elements in Army force projection operations and support endurance and reach operations. They provide loading, unloading, and handling of materiel, cargo, and personnel between various transportation modes. When linked by the modes of transport (air, rail, and sea), they define the physical network for distribution operations. Successful force projection operations require early identification and establishment of terminals.

5-39. Well established terminal operations are essential in supporting deployment, reception, staging, and onward movement and sustainment of the force. Crucial to the successful execution of the terminal/nodal operation is the assignment of the right personnel, cargo, and material handling equipment at each terminal. ITV of materiel moving through the transportation system also provides the GCC with information pertaining to location and destination of all cargo and equipment. There are two types of terminal/nodal operations: marine and inland.

Marine Terminals

5-40. The type, size, number, and location of military marine terminals selected for use, dictate the number and types of units needed to sustain theater support requirements. Using small or geographically dispersed terminals may be necessary for flexibility and survivability. However, this creates a greater need for C2 organizations. A fixed-port facility operated by a HN under contract may only require a contract supervision team. A similar facility operated as a military marine terminal may require a terminal battalion. Fixed-port facilities are designed for oceangoing vessel discharge operations and port clearance. These facilities have sufficient water depth and pier length to accommodate deep-draft vessels. They also have highly sophisticated facilities, equipment, and organization to effectively support cargo discharge and port clearance operations (see FM 55-60). Marine terminals consist of three types of facilities :

- **Fixed-port facilities.** Fixed-port terminals are an improved network of cargo-handling facilities designed to transfer oceangoing freight. These terminals are located worldwide. At these facilities, deep-draft oceangoing vessels berth themselves along a pier or quay and discharge cargo directly onto the pier. Most cargo is moved into in-transit storage areas to await terminal clearance. Selected cargo may be discharged directly to land transport.
- **Unimproved port facilities.** The predominant characteristics of an unimproved port facility are insufficient water depth, insufficient pier length to accommodate oceangoing cargo vessels, and inadequate clearance network. As a result, shallow-draft lighterage must be used to discharge oceangoing vessels anchored in the stream. Other facilities may be available, but they are generally inadequate for cargo discharge operations on a scale associated with a fixed port. In most instances, U.S. Army terminal service units using equipment organic to their TOE operate unimproved port facilities. These terminals are established or used when developed fixed-port facilities are not available or are inadequate to support the workload (see FM 55-60).
- **Bare beach facilities**. A bare beach facility best fits the perceived definition of a LOTS operation. In a bare beach facility, Army lighterage is discharged across a beach. Normally no facilities, equipment, or infrastructure are available to support cargo loading, discharge, or port clearance operations. The terminal service and watercraft units must rely exclusively on equipment organic to their TOE or from supporting engineer units (see FM 55-60).

Inland Terminals

5-41. Inland terminals provide cargo transfer facilities. These include air, motor transport, inland waterway, and rail:

- **Air.** Air cargo transfer takes place at common-use APODs and service controlled airfields and landing strips throughout the theater. A capability assessment should be conducted for each airfield to determine the maximum aircraft on the ground (MOG) that can be parked (called parking MOG) and the number and type of aircraft that can be worked (called working MOG) with available personnel, MHE, and ramp space.
- **Motor transport.** Distribution terminals of Centralized Receiving and Shipping Points are normally located at both ends of a line-haul operation. They form the connecting link between local hauls and the line-haul service. They may also be located at intermediate points along the line-haul route where terrain necessitates a change in type of carrier.
- **Inland waterway.** Inland waterway terminals are limited by the size and configuration of the terminal, types of watercraft, and capabilities of the unit's cargo handling equipment.

- **Rail.** Rail terminals may include yard tracks, repair and servicing facilities, train crew accommodations, and railheads. They are located at originating and terminating points and at sites that mark the limits of rail operations. A railhead can be any size yard or terminal on or at the forward end of a military railway where personnel, supplies, and equipment are transferred to other modes of transportation.

Container Management

5-42. USTRANSCOM has designated SDDC as the global container manager (GCM), to include the authority over execution of container policy across Services as coordinated with a GCC's concept of operations and support. SDDC issues, numbers, and maintains the register of all USAMC DOD-owned inter-modal containers and ISO-configured shelters by DOD Activity Address Code and type container (see FM 4-01.52).

5-43. SDDC provides a theater container database that monitors the inventory, management, and accountability of all containers via the Container Management Element. It also uses automated information systems to monitor movement of containers throughout the theater. The GCM manages, monitors, reports, and provides asset visibility of DOD-owned, leased, and commercial inter-modal surface shipping platforms and containers while in the Defense Transportation System. They provide data expertise to the Army for determining container and container handling equipment requirements to support Army and joint forces contingency, exercises, and peacetime operations.

Containerization

5-44. Containerization facilitates and optimizes cargo carrying capabilities via multiple modes of transport (sea, highway, rail, and air) without intermediate handling of the container's contents. This decrease in time, MHE, personnel, and handling ensures rapid deployment and cargo integrity during shipment. The standardization of the container has facilitated the ease of handling associated with the ISO container and associated MHE. This method of cargo distribution provides fast and flexible preparation, employment, deployment, and sustainment of forces in a theater of operations and extends operational reach. Containerization provides minimum obstructions to the deployment throughput and facilitates unit integrity and cargo security, while enabling container tracking and cargo ITV.

5-45. Service components must plan for theater reception, staging, onward movement, and integration plans. They must include in their plans, methods for container and pallet management and control. When planning to use DOD-owned, Service-owned, or leased containers, the following factors must be considered:

- Availability and location of containers.
- Time and resources required.
- Original out load capability.
- Theater infrastructure/Force structure.
- Availability of MHE at shipping point and at destination.
- Tracking capability, labeling and marking of owner/addressee and destination.
- Method of securing container (lock or serial band).

5-46. A critical node for containerized cargo is the initial entry container control site. This site may be a sea port, aerial port, or rail head. Another critical area of cargo transfer ashore is during LOTS, the shore operations at the beach. At both locations, the container control site will receive, identify, and direct inland distribution and retrograde of containers. The ability of control site personnel to rapidly identify the sender address and the receiver address is the primary enabler of a rapid and successful mission for the force.

5-47. To avoid having large quantities of government-owned containers on hand, the strategy requires partnering with commercial container leasing companies to ensure that leased containers are made available to support military missions which are then staged at depots and power projection platforms in accordance with specific timelines. As operations stabilize, a trans-load operation can commence when directed by the Joint Task Force or CCDR. This would allow government-owned and government-leased containers being used for storage to remain in place, while using ocean carrier provided containers to resupply and sustain operations. These carrier-provided containers would then be unloaded and returned to the carrier within the allotted free time. It is anticipated that these containers will also be used for retrograde operations. Deployment, redeployment, and force rotation requirements will continually be met primarily with government-owned or government-leased containers.

5-48. Current operations resulted in containers being used for non-traditional/non-commercial transportation purposes, such as long-term storage. Future contingencies are expected to be characterized by similar indefinite durations in austere environments. The commercial practice of using ocean carrier provided containers that must be unloaded and returned within a specific time period does not support combat operations and results in high detention costs. Container usage in future contingencies must be addressed in the planning stage in order to minimize cost while supporting mission requirements.

MAINTENANCE

5-49. Maintenance is performed at the tactical through strategic levels of war. The Army's two levels of maintenance are field maintenance and sustainment maintenance (see FM 4-30.3).

FUNDAMENTALS OF ARMY MAINTENANCE

5-50. Field maintenance is repair and return to user and is generally characterized by on-(or near) system maintenance, often utilizing line replaceable unit, component replacement, battle damage assessment, repair, and recovery. Field level maintenance is not limited to remove and replace but also provides adjustment, alignment, and fault/failure diagnoses. Included in field maintenance is the scheduled service/condition based maintenance required on equipment in accordance with the specified technical manual, to include preventative maintenance checks and services. Field maintenance is performed at all levels of the Army and most units have at least some organic field level maintenance capability. Sustainment maintenance is characterized by "off system" component repair and/or "repair and return to supply system." The sustainment maintenance function can be employed at any point in the integrated logistics chain. The intent of this level is to perform commodity-oriented repairs on all supported items to one standard that provides a consistent and measurable level of reliability.

Field Maintenance

5-51. Field maintenance is focused on returning a system to an operational status. The field maintenance level accomplishes this mission by fault isolating and replacing the failed component, assembly, or module. The field maintenance level consists of the maintenance functions of inspection, test, service, adjust, align, remove, replace, and repair. Field maintenance also includes battlefield damage and repair tasks performed by either the crew or support personnel to maintain system in an operational state.

5-52. Within the BCT, the Field Maintenance Company of the BSB provides:

- Automotive, armament, recovery, ground support, missile and electronic maintenance, and maintenance to brigade base elements (HQ, BSB, and Special Troops Battalion).
- Maintenance advice and management to the brigade.
- Low density equipment support to the field maintenance platoon (FMP).

5-53. In the BCT, each maneuver battalion will have a forward support company (FSC) that performs field maintenance. Each FSC has a maintenance platoon that provides recovery support, automotive and tracked vehicle repair, and ground support equipment repair to the battalion. Field maintenance teams deploy with each maneuver company and provide automotive and track vehicle repair support. Ground-support equipment repairs are conducted at the FSC located with the maneuver battalion HQ. The FMP is organized with a maintenance control section to provide maintenance management for the battalion. In EAB, the support maintenance company (SMC) of the Sust Bde provides field maintenance support on an area basis (see FM 4-30.3).

Sustainment Maintenance

5-54. **Sustainment maintenance is generally characterized as "off system" and "repair rear".** The intent of this level is to perform commodity-oriented repairs on all supported items to one standard that provides a consistent and measurable level of reliability. Off-system maintenance consists of overhaul and remanufacturing activities designed to return components, modules, assemblies, and end items to the supply system or to units, resulting in extended or improved operational life expectancies.

5-55. In sustainment maintenance, component repair work is coordinated by the USAMC National Maintenance Office to a single standard that provides consistent and measurable level of reliability. End item repair may be performed by either military or civilian technicians at a sustainment maintenance activity. Repair rear is synonymous with the term "off-system" and "sustainment maintenance". The intent is to repair components, assemblies, or end items and return them to the supply system for redistribution.

5-56. Component repair companies (CRC) provide sustainment level support at the operational level. The CRC may be attached to the CSSB to facilitate overall maintenance support. CRC units may be employed in any location along the distribution system. These units can be pushed forward into the AO as needed to repair and return components, modules, and assemblies to the supply system.

5-57. Collection and Classification companies establish and operate collection and classification facilities for the receipt, inspection, segregation, disassembly, preservation, and disposition of serviceable and unserviceable Class VII and Class IX materiel and similar foreign materiel. It also operates a cannibalization point when authorized by higher HQ. It supports distribution hub teams that perform vital maintenance inspection functions at distribution hubs along the distribution system.

GENERAL ENGINEERING SUPPORT

5-58. General engineering activities modify, maintain, or protect the physical environment (see FM 3-34.400).

Fundamentals of General Engineering

5-59. General engineering capabilities are applied to establish and maintain the infrastructure necessary for sustaining military operations in theater. At times, the military operation may extend general engineering support to restore facilities, power, and life-support systems within the infrastructure of the AO. This effort aids in the recovery and the transition to pre-conflict conditions or may be the objective of stability or civil support operations (see FM 3-34 and FM 3-34.400).

5-60. General engineering capabilities employed in an operation will include a broad array of joint, multinational, contract, and other construction and engineering resources. The U.S. Army Corps of Engineers (USACE) provides and coordinates significant engineering resources to enable general engineering support. USACE is the Army's Direct Reporting Unit assigned responsibility to execute Army and DOD military construction, real estate acquisition, and development of the nation's infrastructure through the civil works program. USACE, through its field force engineering (FFE) and reach back assets, provides for technical and contract engineering support and a means to integrate capabilities of other Services and other sources of engineering-related support.

General Engineering in Support of Sustainment Operations

5-61. General engineering sustainment requirements will compete for priority in any operation with general engineering requirements related to protection, enabling operational movement, as augmentation to combat engineering, and supporting the other WFFs. Within the sustainment WFF, general engineer applications are primarily linked to providing logistics support. General engineer support includes:

- Restore Damaged Areas.
- Construct and Maintain Sustainment Lines of Communications.
- Provide Engineer Construction (to include pipeline) Support.
- Supply Mobile Electric Power.
- Provide Facilities Engineering Support.

5-62. Sustainment of stability and civil support operations involves a shift to the establishment of services that support civilian agencies in addition to the normal support of U.S. forces. Stability operations tend to be of a long duration compared to the other elements in full spectrum operations. As such, the general engineering level of effort, including FFE support from USACE, is very high at the onset and gradually decreases as the theater matures. As the AO matures, the general engineering effort may transfer to theater or external support contracts such as LOGCAP.

ENGINEER COORDINATION

5-63. The senior engineer staff officer at each echelon HQ, designated the ENCOORD, is responsible for coordinating sustainment related general engineer support. An engineer brigade or theater engineer command will typically be task organized with those general engineering capabilities not provided to subordinate BCTs or other brigades. The engineer brigade or theater engineer command focuses general engineering efforts on priorities established by the JFC.

5-64. Engineering priorities will typically include sustainment related general engineer support. The engineer support commander may align engineer assets to provide general support on an area basis. If assets are available and priorities support a more direct relationship, the commander may place an engineer brigade in DS to the TSC, with subordinate engineer elements DS to the support brigades or CSSBs as required.

REAL ESTATE PLANNING AND ACQUISITION

5-65. The JFC is responsible for the coordination of planning, programming, and construction of facilities to meet the requirements of assigned forces. Facility requirements are consistent with operational requirements, duration of need, and forces to be supported. Engineer planners coordinate with sustainment and other planners to identify facility requirements for contingency operations.

5-66. USACE theater elements provide technical real estate guidance and advice to the JFC. In addition to recommending real estate policies and operational procedures, they acquire, manage, dispose of, administer payment for rent and damages, handle claims, and prepare records and reports for real estate used within the theater.

REAL PROPERTY MAINTENANCE

5-67. The JFC has overall responsibility for real property maintenance activities (RPMA). The JFC normally delegates authority to the ASCC/ARFOR. The TSC and installation commanders (in most cases a CSSB) normally provide the needed RPMA support. RPMA in an AO includes operation, repair, and maintenance of facilities and utilities; fire prevention and protection; and refuse collection and disposal. RPMA requirements that exceed the organization's capabilities are forwarded to the local engineer commander (in most cases, the engineer group providing support to a CSSB on an area basis) or USACE element for execution. The TSC provides technical RPMA guidance to subordinate units.

BASE CAMP CONSTRUCTION

5-68. A base camp is an evolving military facility that supports the military operations of a deployed unit and provides the necessary support and services for sustained operations. It is a grouping of facilities collocated within a contiguous area of land, or within close proximity to each other, for the purpose of supporting an assigned mission, be it tactical, operational, or logistical. Base camps may be located near a key piece of real estate such as a port, an airfield, a railroad, or other major LOCs. Base camps support the tenants and their equipment; and while they are not installations, they have many of the same facilities and attributes the longer they are in existence.

5-69. The CCDR specifies the construction standards for facilities in the theater to minimize the engineer effort expended on any given facility while assuring that the facilities are adequate for health, safety, and mission accomplishment. Typically, the CCDR will develop the base camp construction standards for use within the theater, utilizing the guidelines provided in JP 3-34 and facilities standards handbooks developed by the specific combatant command. The engineer must recommend the most feasible solutions to each requirement based on construction guidelines and other planning factors.

ENVIRONMENTAL CONSIDERATIONS

5-70. The ENCOORD also advises the commander on environmental issues as the staff proponent for environmental considerations. The ENCOORD coordinates with other staff offices to determine the impact of operations on the environment and helps the commander integrate environmental considerations into the decision making process. Environmental considerations include:

- Policies and responsibilities to protect and preserve the environment.
- Certification of local water sources by appropriate medical personnel.
- Solid, liquid, and hazardous waste management, including dumping and burning, and disposal of gray water, pesticides, human waste, and hazardous materials.
- Protection of indigenous animals and vegetables.
- Archaeological and historical preservation.
- Contingency spill plans.

HUMAN RESOURCES SUPPORT

5-71. HRS is the aggregate of systems and services designed to provide and support Soldiers. HRS is important to maximizing operational reach and endurance. HRS encompasses four major categories: manning the force, HR services, personnel support, and HR planning and staff operations. Each includes major functional elements and all are covered below (see FM 1-0, FMI 1-0.1, and FMI 1-0.2).

MANNING THE FORCE

5-72. Manning the force involves personnel readiness of the force, maintaining accountability of the force, and management of personnel information. The manning challenge is getting the right Soldier to the right place, at the right time, with the right capabilities so that commanders have the required personnel to accomplish their mission. Manning combines anticipation, movement, and skillful positioning of personnel assets. It relies on the secure, robust, and survivable communications and digital information systems of emerging technologies that provide the common operational picture, asset visibility, predictive modeling, and exception reporting.

Personnel Readiness Management (PRM)

5-73. The purpose of PRM is to distribute Soldiers to units based on documented requirements, authorizations, and predictive analysis to maximize mission preparedness and provide the manpower needed to support full spectrum operations. This process involves analyzing personnel strength data to determine current mission capabilities and project future requirements. It compares an organization's personnel strength to its requirements and results in a personnel readiness assessment and allocation decision.

Personnel Accountability (PA)

5-74. PA plays a critical role in deployed operations and relies on timely, accurate, and complete duty status and location of personnel at all times. PA is the process for recording by-name data on Soldiers when they arrive and depart from units; when their location or duty status changes (such as from duty to hospital); or when their grade changes. PA will be accomplished primarily through the database of record and web enabled processes that facilitate personnel support from home station or abroad. Personnel Accounting Teams manage or administer all HR support activities of processing, tracking, and coordinating personnel moves into, through, or out of a deployed organization or theater. These activities include the reception of personnel, the assignment and tracking of replacements, return-to-duty and rest and recuperation personnel, and redeployment operations.

Strength Reporting

5-75. Strength reporting is a numerical end product of the accounting process, achieved by comparing the by-name data obtained during the personnel accountability process (faces) against specified authorizations (spaces or in some cases requirements) to determine a percentage of fill. It starts with strength-related data submitted at unit level and ends with an updated database visible at all echelons. Similar to PA, strength reporting relies on timely, accurate, and complete personnel information into the database of record. It is also a command function conducted by the G-1/S-1 to enable them to provide a method of measuring the effectiveness of combat power. Standard reports available from the personnel accounting system include:

- Personnel status report.
- Personnel summary.
- Personnel requirements report.
- Task force personnel summary.

Personnel Information Management (PIM)

5-76. PIM encompasses the collecting, processing, storing, displaying, and disseminating of relevant information about Soldiers, units, and civilians. PIM is the foundation for conducting or executing all HR functions and tasks. HR managers and technicians at all levels of command use a personnel information database when performing their missions. The DIMHRS, when implemented, will be the HR enterprise database for all military personnel.

PROVIDE HR SERVICES

5-77. HR Services encompass casualty operations and EPS to maintain Soldier readiness and sustain the human dimension of the force. The following is a discussion of casualty operations, EPS, and the elements of personnel support.

Casualty Operations Management

5-78. The casualty operations management process includes the recording, reporting, verifying, and processing of information from unit level to HQ, Department of the Army. It also involves notifying appropriate individuals and assisting family members. The process collects casualty information from a number of sources and then collates, analyzes, and determines the appropriate action. Accuracy and timeliness are critical components of casualty management and depend on satellite communications and reliable access to personnel information.

5-79. Casualties can occur on the first day of an operation. Therefore, casualty managers from each echelon of command need to deploy early. Units report all casualties, to include civilians, contract, and military personnel from Army, other services, and MNFs. Casualty operations require 100 percent personnel accounting reconciliation. The unit verifies casualty information against the database and emergency data in an individual's deployment packet. Casualty liaison teams (CLT) provide an interface between medical facilities, MACP collection points, and human resources elements.

Essential Personnel Services (EPS)

5-80. EPS provide Soldiers and units timely and accurate personnel services that efficiently update Soldier status, readiness and quality of life, and allow Army leadership to effectively manage the force. EPS includes actions supporting individual career advancement and development, proper identification documents for security and benefits entitlements, recognition of achievements, and service performance. It also includes personal actions such as promotions, reductions, evaluations, military pay, leave and pass, separations, and line-of-duty investigations.

PERSONNEL SUPPORT

5-81. Personnel Support encompasses command interest/human resources programs, MWR, and retention functions. Personnel Support also includes substance abuse and prevention programs, enhances unit cohesion, and sustains the morale of the force.

Postal Operations

5-82. Postal operations and services have a significant impact on Soldiers, civilians, and their families. The Military Postal Service serves as an extension of the U.S. Postal Services; therefore, its services are regulated by public law and federal regulation. Efficient postal operations are necessary and require significant logistics and planning for issues such as air and ground transportation, specialized equipment, secured facilities, palletization crews, mail handlers, and others. Postal services also include selling stamps; cashing and selling money orders; providing registered (including classified up to secret), insured and certified mail services; and processing postal claims and inquiries.

Morale, Welfare, and Recreation (MWR) and Community Support

5-83. MWR and community support provide Soldiers, Army civilians, and other authorized personnel with recreational and fitness activities, goods, and services. The MWR network provides unit recreation, library books, sports programs, and rest areas for brigade-sized and larger units. Community support programs include the American Red Cross, AAFES, and family support system.

5-84. The MWR system becomes an immediate outlet for Soldiers to reduce stress, which is critical to sustaining the readiness of the force, particularly as the speed and intensity of operations escalate. The MWR system relies on FP packages and recreation specialists. It capitalizes on using cellular, e-mail, and video-teleconference technologies to provide links between Soldiers and their families. Soldiers are also entertained through the latest in visual and audio entertainment over satellite, worldwide web, and virtual reality technologies. The human dimension of the Soldier is critical to the strength of Army forces. The human resource element of sustainment to the fighting force contributes to both the National will and the will of the Soldier to fight.

Human Resources Planning and Staff Operations

5-85. Human Resources Planning and Staff Operations are the means by which the HR provider envisions a desired HR end state in support of the operational commander's mission requirements. HR planning addresses the effective ways of achieving success, communicates to subordinate HR providers and HR unit leaders the intent, expected requirements, and outcomes to be achieved, and provides the support OPLANs, OPORDs, or Planning Annex.

5-86. Planning and staff operations are also the process of tracking current and near-term (future) execution of the planned HR support to ensure effective support to the operational commander through the military decision process. Effective planning includes ensuring HR C2 nodes are established, operated, and that connectivity to HR data and voice communications nodes is maintained. HR C2 nodes include those required for all HR operations, across commands and echelons, and to higher and lower elements.

FINANCIAL MANAGEMENT (FM)

5-87. The FM mission is to analyze the commander's tasks and priorities to ensure that proper financial resources are available to accomplish the mission and to provide recommendations to the commander on the best allocation of scarce resources. FM support enhances the commander's ability to manage and apply available resources at the right time and place in a fiscally responsible manner. FM provides the capability for full spectrum finance and RM operations across the theater to include all unified operations.

5-88. FM is comprised of two core functions: finance operations and RM. These two processes are similar and mutually supporting in organizational structure and focus. The ASCC G-8, in consultation with the FMC Director, is the principal adviser to the ASCC in all matters relating to FM operations. The integration of finance and RM under one entity offers the commander a single focal point for FM operations (see FM 1-06).

FINANCE OPERATIONS

5-89. Finance operations must be responsive to the demands of the unit commanders at all levels, requiring FM leaders to anticipate and initiate the finance support needed. This section summarizes finance operations during all operational stages. It presents a stable body of technical principles rooted in actual military experience from which commanders can guide their actions in support of national objectives.

Procurement Support

5-90. The success of all operations depends on the support provided to the sustainment system and to contingency contracting efforts. A large percentage of the FM mission is to support the procurement process and provide oversight. Oversight is critical in preventing improper or illegal payments. By coordinating with the contracting officer and the SJA regarding local business practices, financial managers greatly reduce the probability of improper or illegal payments. Procurement support includes two areas: contracting support and commercial vendor services (CVS) support.

5-91. Contracting support involves payment to vendors for goods and services. This includes all classes of supply, laundry operations, bath operations, transportation, and maintenance. Financial managers are crucial to successful contracting operations.

5-92. CVS provides for the immediate needs of the force. These are needs the standard logistics systems cannot support. This usually includes payments of cash (U.S. or local currency). Cash payments are usually for day laborers, Class I supplements (not otherwise on contract) and the purchase of construction material not readily available through the contract or supply system.

Limited Pay Support

5-93. FM units provide limited travel support, casual payments, check cashing and currency exchanges to Soldiers and civilians in permanent change of station and temporary duty status, NEO travel advances, and non-US pay support (EPW, civilian internee, host nation employees, and day laborers).

Disbursing Support

5-94. Disbursing support includes training and funding paying agents in support of local procurement, administering the Stored Value Card (SVC), supporting rewards programs, and making condolence and solatium payments. Individual support is provided to Soldiers and/or civilians through check cashing, foreign currency conversions, receiving collections (such as Savings Deposit Program), making payments on prepared and certified vouchers, funding FM units, determining the need for currency (U.S. and foreign) and its replenishment, and receiving and processing all captured currencies and precious metals.

Accounting Support

5-95. Accounting support includes ensuring proper financial resources are available to the commander by supporting the fiscal triad (Contracting, RM, and Finance) in reconciling expenditures and thus providing the most accurate and timely financial data.

Banking and Currency Support

5-96. Banking relationships and procedures are established with any banking industry of a host nation to include establishing local depository accounts, limited depository accounts for current contract payments, and foreign currency resupply.

Resource Management Support

5-97. RM operations are a critical enabler at all levels of the Army's chain of command. The RM operations mission is to analyze tasks and priorities and to identify the resource requirements that will enable the commander to accomplish the mission. In advising the commander, financial managers perform the tasks shown below.

- Identify, Acquire, Distribute and Control Funds. Financial managers identify the sources of funds available from various DOD and other Federal agencies. They also acquire the funds and distribute funds to subordinate elements to support the mission and commander's intent.
- Develop Resource Requirements. Determining what resources are required and available to support the mission and commander's intent includes, but is not limited to, contracting, transportation, multinational support, support to other agencies and international organizations,

foreign humanitarian assistance, and force sustainment. Developing and determining resource requirements also includes:

- Preparing the FM annex to operations plan and order.
- Determining and validating costs to accomplish the mission.
- Determining when resources are needed throughout the fiscal year(s).
- Making resources available at the time and amount needed.
- Developing budgets.
- Coordinating fiscal issues associated with all unified action operations.

- Track, Analyze, and Report Budget Execution. Procedures are established to track costs in order to determine obligation rates and conduct analysis on use of funds in support of the mission and to identify trends to foresee resourcing challenges. Reports are submitted as required by policy. Tracking, analyzing, and reporting budget execution include the following:
  - Analyze RM and accounting reports.
  - Establish procedures to track costs.
  - Establish management internal control processes.
- Accounting Support. Financial managers ensure official accounting records are accurate, properly supported by source documentation, and resolve accounting issues in a timely manner.

5-98. The Army may be appointed the EA responsible for FM operations. The EA for FM normally will fund multi-Service contract costs, unique joint force operational costs, special programs, joint force HQ operational costs, and any other designated support costs. The EA also provides financial analysis and recommendations to joint forces for the most efficient use of fiscal resources (see JP 1-06, Financial Management Support in Joint Operations, for further information).

5-99. Regardless of the scale or scope of sustainment operations, finance and RM operations play a key role in providing responsive agile support to deployed forces across the spectrum of conflict. Each of these operations must be fully integrated and synchronized with all other facets of sustainment operations in order to effectively and efficiently sustain the force (see FM 1-06).

LEGAL SUPPORT

5-100. Members of The Judge Advocate General's Corps (JAGC) provide proactive legal support on all issues affecting the Army and the Joint Force and deliver quality legal services to Soldiers, retirees, and their families. Legal support centers on six core disciplines across full-spectrum operations. The six core disciplines are: military justice, international and operational law, contracts and fiscal law, administrative and civil law, claims, and legal assistance. Each discipline is described below (see FM 1-04).

MILITARY JUSTICE

5-101. Military justice is the administration of the Uniform Code of Military Justice (UCMJ). The purpose of military justice, as a part of military law, is "to promote justice, to assist in maintaining good order and discipline in the armed forces, to promote efficiency and effectiveness in the military establishment, and thereby to strengthen the national security of the United States." (Preamble, *Manual for Courts-Martial* (MCM) (2008)). The Judge Advocate General is responsible for the overall supervision and administration of military justice within the Army. Commanders are responsible for the administration of military justice in their units and must communicate directly with their servicing SJAs about military justice matters (AR 27-10).

INTERNATIONAL AND OPERATIONAL LAW

5-102. International law is the application of international agreements, U.S. and international law, and customs related to military operations and activities. Within the Army, the practice of international law includes the interpretation and application of foreign law, comparative law, martial law, and domestic law affecting overseas activities, intelligence, security assistance, counter-drug, stability operations, and rule of law activities.

5-103. Operational Law (OPLAW) is that body of domestic, foreign, and international law that directly affects the conduct of military operations. OPLAW encompasses the law of war, but goes beyond the traditional international law concerns to incorporate all relevant aspects of military law that affect the conduct of operations. The OPLAW attorney supports the commander's military decision-making process by performing mission analysis, preparing legal estimates, designing the operational legal support architecture, war gaming, writing legal annexes, assisting in the development and training of rules of engagement (ROE), and reviewing plans and orders. The OPLAW attorney supports the conduct of operations by maintaining situational awareness and advising and assisting with targeting, ROE implementation, and information tasks. OPLAW also involves the provision of core legal disciplines that sustain the force.

Administrative and Civil Law

5-104. Administrative law is the body of law containing the statutes, regulations, and judicial decisions that govern the establishment, functioning, and command of military organizations. The practice of administrative law includes advice to commanders and litigation on behalf of the Army involving many specialized legal areas, including military personnel law, government information practices, investigations, relationships with private organizations, labor relations, civilian employment law, military installations, regulatory law, intellectual property law, and government ethics.

5-105. Civil law is the body of law containing the statutes, regulations, and judicial decisions that govern the rights and duties of military organizations and installations with regard to civil authorities. The practice of civil law includes environmental law, as well as other specialized areas of law.

Contract and Fiscal Law

5-106. Contract law is the application of domestic and international law to the acquisition of goods, services, and construction. The practice of contract law includes battlefield acquisition, contingency contracting, bid protests and contract dispute litigation, procurement fraud oversight, commercial activities, and acquisition and cross-servicing agreements. The SJA's contract law responsibilities include furnishing legal advice and assistance to procurement officials during all phases of the contracting process, overseeing an effective procurement fraud abatement program, and providing legal advice to the command concerning battlefield acquisition, contingency contracting, use of LOGCAP, ACSAs, the commercial activities program, and overseas real estate and construction.

5-107. Fiscal law is the application of domestic statutes and regulations to the funding of military operations and support to non-federal agencies and organizations. Fiscal law applies to the method of paying for obligations created by procurements.

Claims

5-108. The Army Claims Program investigates, processes, adjudicates, and settles claims on behalf of and against the United States world-wide under the authority conferred by statutes, regulations, international and interagency agreements, and DOD Directives. The Claims Program supports commanders by preventing distractions to the operation from claimants, promoting the morale of Army personnel by compensating them for property damage suffered incident to service, and promoting good will with the local population by providing compensation for personal injury or property damage caused by Army or DOD personnel.

Legal Assistance

5-109. Legal assistance is the provision of personal civil legal services to Soldiers, their family members, and other eligible personnel. The mission of the Army Legal Assistance Program is to assist those eligible for legal assistance with their personal legal affairs in a timely and professional manner by meeting their personal needs for response and information on legal matters and then resolving the legal problems whenever possible. From an operational standpoint, the provision of legal services at the earliest possible time is critical to ensure the readiness of individual Soldiers and the force as a whole.

PROVISION OF LEGAL SUPPORT

5-110. Units at the BCT level and echelons above have organic legal elements to support the mission. At the BCT, the brigade legal section (BLS) is responsible for providing services in all core legal disciplines that extend across the full spectrum of operations. The BLS mission is dictated primarily by the brigade commander's guidance and direction and the brigade judge advocate's professional judgment. The level of service that a BLS is able to provide will depend on a number of factors, including: the brigade's operational tempo, the brigade's deployment status, the experience level of the BLS, the availability of additional judge advocate or paralegal support during "surge" periods, and the existence of actual conflicts of interest. When faced with situations where the BLS is unable to provide the proper level of service, the brigade judge advocate should use the brigade chain of command and the JAGC technical channels to address the shortfalls.

5-111. Units at division level and higher receive legal support from an Office of the Staff Judge Advocate which is responsible for the provision of legal services across all core legal disciplines to the appropriate commander and General Court Martial Convening Authority. The division SJA is a personal staff officer with direct access to the commander. In addition, SJAs typically possess logistical and professional capabilities which allow them to enhance the resources of a subordinate legal section.

RELIGIOUS SUPPORT

5-112. Religious support facilitates the free exercise of religion, provides religious activities, and advises commands on matters of morals and morale. The First Amendment of the U.S. Constitution and Army Regulation (AR) 165-1 guarantees every American the right to the free exercise of religion. Commanders are responsible for fostering religious freedoms. Chaplains and chaplain assistants functioning as Unit Ministry Teams (UMT) perform and provide RS in the Army to ensure the free exercise of religion (see FM 1-05).

FUNDAMENTALS OF RELIGIOUS SUPPORT

5-113. RS to the Army is guided by historical precedence. The three broad functions of religious support include nurturing the living, caring for the wounded, and honoring the dead. These include all other specific activities carried out by chaplains and chaplain's assistants, as discussed below.

Nurture the Living

5-114. In preparation for missions that span the spectrum of conflict, UMTs develop and provide religious support activities to strengthen and sustain the spiritual resilience of Soldiers and family members. During the operation, UMTs bring hope and strength to those who have been wounded and traumatized in body, mind, and spirit, thus assisting the healing process.

Caring for the Wounded

5-115. UMTs provide religious support, spiritual care, comfort, and hope to the wounded. This focus of RS affirms the sanctity of life, which is at the heart of the chaplaincy. Through prayer and presence, the UMT provides the Soldiers and their families with courage and comfort in the face of death.

Honoring the Dead

5-116. Our Nation reveres those who have died in military service. RS honors the dead. Funerals, memorial ceremonies, and services reflect the emphasis the American people place on the worth and value of the individual. Chaplains conduct ceremonies and services, fulfilling a vital role in rendering tribute to America's sons and daughters who paid the ultimate price serving the Nation in the defense of freedom.

KEY ASPECTS TO RELIGIOUS SUPPORT

5-117. There are several additional aspects of religious support. They include:

- Facilitating individual freedom of worship and observation of holy days in accordance with Army regulations and mission requirements.

- Advising the command on morals and morale as affected by religion and the impact of indigenous religions.
- Advising the command on the ethical impact of command decisions, policies, and procedures.
- Resolution of medical treatment religious and ethical issues, religious apparel issues, and religious dietary restrictions in accordance with AR 600-20.
- Respect for the constitutional, statutory, and regulatory requirements ensuring freedom of religion for every Soldier, family member, and command authorized civilian.

RELIGIOUS SERVICES

5-118. Chaplains are obligated to provide for those religious services or practices that they cannot personally perform. Chaplains perform religious services when their actions are in accordance with the tenets or beliefs of their faith group. Chaplain assistants assist the chaplain in providing or performing this religious support.

- Unit Support is provided to the unit to which the UMT is assigned or attached. The team normally gives priority to this mission.
- Area Support is provided to Soldiers, members of other services, and authorized civilians who are not a part of the team's unit, but operate within the unit's AO.
- Denominational Support is given to Soldiers and other authorized persons of the chaplain's denomination or distinctive faith group. Availability of assets may limit the availability of denominational support provided.

STAFF INTEGRATION AND COORDINATION

5-119. Chaplains personally deliver religious support. They have two technical roles: religious leader and religious staff advisor. The chaplain as a religious leader executes the religious support mission which ensures the free exercise of religion for Soldiers and authorized personnel. The chaplain is a non-combatant and will not bear arms (see AR 165-1).

5-120. Chaplain assistants are enlisted personnel and are combatants. As combatants, they integrate UMTs into tactical formations for security and survivability. Chaplain assistants also possess specific technical and staff competencies to support administering the Command Master Religious Program on behalf of the commander.

5-121. The chaplain is a personal staff officer responsible for coordinating the religious support assets and activities within the command. The chaplain is a confidential advisor to the commander for religious matters. A chaplain is located at every echelon of command from battalion through Army Service component command (see FM 1-05).

BAND SUPPORT

5-122. Army bands provide critical support to the force by tailoring music support throughout military operations. Music instills in Soldiers the will to fight and win, foster the support of our citizens, and promote America's interests at home and abroad. (see FM 1-0 and FM 1-19)

FUNDAMENTALS OF BAND SUPPORT

5-123. Army bands sustain the operational Army throughout the full spectrum of operations through the provision of tailored music that enhances Warrior morale, supports Army recruiting efforts, and maintains a connection with the American public. Music serves as a useful tool to reinforce relations with host nation populations and favorably shapes the civil situation throughout the peace building process. Inherently capable of providing a climate for international relations, bands serve as ambassadors in multi-national operations or to the host nation population.

5-124. Army bands are modular units designed to support Army, Joint, and Multinational formations. Army bands are organized into four types: Small, Medium, Large, and Special. Army Bands Small and Medium are further subdivided based upon their operational capabilities with regard to the deployment in support of ongoing operations. Special bands have unique responsibilities in support of the Military District of Washington, Headquarters, Department of the Army Public Affairs, or the United States Military Academy.

5-125. The modular structure of Army bands, with "plug-in" augmentation, enables split operations in support of musical mission requirements. Home station missions among Army bands include music support of Soldier and family, wounded warrior outreach, and community relations. Deployment operations of Army bands include the music support of morale-building events among Soldiers, joint-services, and MNFs as well as building alliances or shaping the civil situation with host nation and/or regional populations. Army bands can also provide music support of nondenominational religious activity in accordance with AR 220-90, both the at home station and during deployment operations.

HEALTH SERVICE SUPPORT

5-126. The AHS is a component of the Military Health System that is responsible for operational management of the HSS and FHP missions for training, pre-deployment, deployment, and post deployment operations. AHS support includes all mission support services performed, provided, or arranged by the AMEDD to support HSS and FHP mission requirements for the Army and as directed, for joint, intergovernmental agencies, coalitions, and MNFs (see FM 4-02).

FUNDAMENTALS OF ARMY HEALTH SERVICES

5-127. AHS support is guided by six fundamentals. These fundamentals are consistent with JP 4-02. They are described below:

- OPLAN conformance. By ensuring that Force Health Projection support conforms to the tactical commander's OPLAN, the AHS support planner can determine support requirements and plan for the support needed to prevent non-battle injuries and to effectively clear the battlefield of the ill, injured, and wounded.
- Surgeon technical direction. Technical direction/guidance and staff supervision of AHS support activities must remain with the appropriate command-level surgeon.
- Continuity of care. The AHS support staff must maintain continuity of care since an interruption of treatment may cause an increase in morbidity and mortality. No patient is evacuated farther to the rear than his/her medical condition or the tactical situation dictate.
- Proximity. The proximity of AHS support assets to the supported forces is dictated by the tactical situation METT-TC.
- Flexibility. The AHS support plan must be flexible to enhance the capability of reallocating AHS support resources to meet changing requirements. Changes in the tactical situation or OPLAN make flexibility essential.
- Mobility. Mobility is required to ensure that AHS support assets remain close enough to combat operations to support combat forces. The mobility and survivability of medical units and medical platforms must be equal to the forces supported.

ARMY HEALTH SYSTEM SUPPORT

5-128. AHS support involves the delineation of support responsibilities by capabilities (roles of care) and geographical area (area support). The AHS support executes its HSS and FHP missions as a single, seamless, and integrated system. The AHS support encompasses the promotion of wellness and preventive, curative, and rehabilitative medical services. It is designed to maintain a healthy and fit force and to conserve the fighting strength of deployed forces. The goal of the AHS in support of the HSS and FHP missions is to:

- Provide prompt medical treatment consisting of those measures necessary to locate, acquire, resuscitate, stabilize, document, and prepare patients for evacuation to the next role of care and/or return to duty (RTD).
- Employ standardized air and ground medical evacuation units/resources, in conjunction with the aviation brigades for air ambulances.

- Provide flexible, responsive, and deployable medical support designed and structured to sustain a force projection Army and its varied missions. This capability includes hospitalization resources to provide essential care to all patients who cannot recover within the theater evacuation policy and are stabilized and evacuated out of theater and definitive care to those Soldiers capable of returning to duty (see FM 4-02.10).
- Provide a medical logistics system (to include blood management) that is anticipatory and tailored to continuously support missions throughout full spectrum operations (see FM 4-02.1).
- Provide dental services to maximize the RTD of dental patients by providing operational dental care and maintaining the dental fitness of theater forces.
- Provide medical laboratory functions in medical operations.
- Provide blood management services.
- Provide preventive dentistry activities.
- Provide combat and operational stress control and behavioral health (BH) preventive services.

SUMMARY

5-129. The functional elements of sustainment include supply, field services, transportation, maintenance, general engineering, human resources, FM, legal, religious support, and Army health services support. These elements and their many sub-functions comprise the sustainment WFF. When optimized, sustainment operations ensure strategic and operational reach and endurance for Army forces in any operational environment.

Appendix A

Information Systems

Information systems are essential for providing commanders and staffs situational understanding and building the common operational picture. This appendix describes several of the C2 and STAMIS systems used in sustainment operations. Highlighted below are key systems. However, the list is not all inclusive.

HUMAN RESOURCES SYSTEMS

A-1. **Defense Integrated Military Human Resources System (DIMHRS).** DIMHRS, when implemented, will be a fully integrated web-based, all-Service, all-Component, military personnel and pay system that will support military personnel throughout their careers. DIMHRS, when fielded, will be the official database of record for military personnel and provides a fully integrated single source joint database responsible for personnel and pay processes. All DIMHRS transactions support the personnel life-cycle functions of Access/Retain, Assign, Sustain, Evaluate/Promote, and Separate. DIMHRS will replace approximately 67 legacy Army systems, integrating payroll and personnel functions for the Regular Army and Reserve Components. See FM 1-0 for additional information on DIMHRS.

A-2. **Defense Casualty Information Processing System (DCIPS).** DCIPS is a single uniform casualty reporting system for use by all services. This system manages receipt of the casualty messages, permits interactive update of casualty information, allows data exchange with Casualty Assistance Centers and mortuaries, the Armed Forces Medical Examiner, the Defense Manpower Data Center, and various other medical surveillance organizations, and formulates required documents and reports. DCIPS interfaces with (DIMHRS) to obtain personnel data. It also interfaces with the U.S. Air Force Mortuary Operations Management System at Dover Air Force Base to exchange the casualty personnel information, identification of remains information, family disposition instructions, and remains shipping information. See FM 1-0 for additional information on DCIPS.

A-3. **Defense Casualty Information Processing System-Forward (DCIPS-Fwd).** DCIPS-Fwd is an automated system to record and report casualty data. The system is employed with HR units performing casualty reporting missions and is capable of producing automated casualty reports. See Chapter 5 for information on Casualty Operations or FM 1-0 for additional information on DCIPS-Fwd.

A-4. **Deployed Theater Accountability Software (DTAS).** DTAS fills the current void within the Personnel Automation Architecture. It provides the essential personnel functionality to support a commander's tactical decision-making process. DTAS is a SIPR-based theater personnel accountability and strength reporting system. DTAS builds a deployed database. This capability is critical for immediate and future operations. DTAS capabilities include:

- Bridge the gap between hasty strength reporting and deliberate personnel accountability.
- Assist S-1/G-1 in personnel accountability, strength reporting, and replacement operations management through reports and queries.
- Update personnel strength when duty status changes are received from units, medical facilities, MA, and MP straggler control.
- Allow HR managers to initiate verification of duty status changes.
- Produce and edits task force structures by unit identification code, social security number, and/or crew.

A-5. **Synchronized Pre-deployment Operational Tracker (SPOT).** SPOT is a web accessible database designed to account for contractor personnel. SPOT also provides a basic level of contract information and contactor personnel accountability. In January 2007, the Office of the Secretary of Defense designated SPOT as the central repository database for all contractor personnel information.

FINANCIAL MANAGEMENT (FM) SYSTEMS

A-6. **General Funds Enterprise Business System (GFEBS).** GFEBS is the Army's core FM system to provide capabilities such as distribution and execution of appropriated funds, cost management, financial reporting, and management of real property.

A-7. **The Corporate Electronic Document Management System (CEDMS).** CEDMS is a web-based electronic file room. It eliminates the expensive and labor intensive requirement of maintaining paper files by providing a centralized repository of digital documents. CEDMS provides a secure, high performance, scalable, and reliable centralized repository that will accommodate the administrative requirements, to include document management, record keeping, record retrieval, record staging, retention, contingency operations, and document security, for scanning, indexing, and managing DFAS documents.

A-8. **Wide Area Work Flow (WAWF).** WAWF is an e-commerce business solution for DOD and defense contractors. It allows online submission of invoices/receiving report and electronic disbursement to vendors. WAWF helps the Army reduce unmatched disbursement through electronic sharing of receipt, acceptance, entitlement, and the payment documents. It eliminates cash and check payment to vendors and reduces costs to Army; no lost paperwork or processing delays.

A-9. **Resource Management Tool (RMT).** RMT consolidates and integrates financial and manpower data from multiple sources into a single database. RMT links unit FM information into the Standard Finance System (STANFINS) and when fully deployed, GFEBS. RMT provides real time execution data and a common solution for manpower distribution, workload forecasting, and performance measurement.

A-10. **International Treasury Services.gov (ITS.gov).** ITS.gov is an international payment and collection system used for processing international direct deposit payments to benefit recipients with both electronic and check payments for vendor pay, foreign payroll, and miscellaneous payment recipients in foreign countries.

A-11. **Cash-Link.** Cash-Link is a web-based system used to research Treasury deposits and debit transactions. Cash-Link reduces the time spent on researching and correcting erroneous or disputed EFT transactions and to reconcile Army accounts with the U.S. Treasury.

A-12. **Paper Check Conversion (PCC).** PCC is a Treasury system which converts a personal/business checks into electronic funds transfers. PCC provides real-time access to customer's checking accounts and eliminates non-sufficient funds transactions and the time it takes to process them.

A-13. **Financial Management Tactical Platform (FMTP).** FMTP is a deployable, modular local area network-configured hardware platform that supports finance and RM operations and functions across the entire spectrum of conflict. System functionality includes vendor services, military pay, disbursing, accounting, travel, and RM. FMTP is a "system wrapper." It integrates essential Defense Finance and Accounting Systems into a user-friendly package. The system operates on non-developmental item hardware at all levels of the deployed environment, wherever FM units can be found. Software available to deployed financial managers include: 1) Deployable Disbursing System (DDS). DDS provides automated accounting and disbursing documentation to mobile and remote military operations within contingency locations. DDS supports operations in remote combat environments where communications capabilities are limited. It replaces labor-intensive manual processes; integrates the flow of information between entitlements and accounting; improves the accountability of funds, timeliness of financial information, and the provision of historical data. DDS reformats data and links it to Federal Reserve Banks (FRB) of NY and ATL. 2) Defense Joint Military Pay System (DJMS). DJMS is the joint system for pay and entitlements for the Active and Reserve Components. 3) Integrated Automated Travel System (IATS). IATS is seamless, paperless temporary duty travel system interfacing with accounting, reservation, disbursing, banking, and archiving systems. 4) Commercial Accounts Payable System (CAPS). CAPS is used to automate manual functions in the accounts payable offices such as: automatically suspense commercial payments and follow-up letters, provide payment computations, produce vouchers and management reports, compute the payment due date, interest penalties and determine lost discounts, allow for entry and processing of purchase rates, purchase orders/ contracts, and determine foreign currency rates, and maintain the Electronic fund transfer information for vendors whose contracts specify this type of payment. 5) The Database Commitment Accounting System (dbCAS). This system provides for the input and tracking of all commitment and obligation information; dbCAS can receive downloaded STANFINS information to assist in matching commitments with obligations and disbursements. It also produces reports that allow the commander to determine current fund status. 6) Paper Check Conversion Over the Counter (PCC OTC). PCC-OTC converts personal and business checks into electronic funds transfers. 7) Eagle Cash SVC. The SVC is a credit card-sized plastic card with an integrated circuit embedded in it. Typically SVC cards have no value until they are activated and have funds placed on them electronically. Value can be added to cards in a number of ways, including payroll or other financial file transfer, from a credit or debit card or from cash or checks.

LOGISTICS SYSTEMS

A-14. **Global Combat Support System Army (GCSS-Army)** GCSS-Army is replacing a variety of legacy tactical-level logistics information systems and automated capabilities such as the Standard Army Retail Supply System (SARSS), the Standard Army Maintenance System-Enhanced (SAMS-E), Unit Level Logistics System Aviation Enhanced (ULLS-AE), and the Property Book Unit Supply Enhanced (PBUSE). The Army Enterprise System Integration Program (AESIP) will link GCSS-Army—the Army's field-level logistics system—with Logistics Modernization Program (LMP)—the Army's national-level logistics system. GCSS-Army will provide a single access point to the Single Army Logistics Enterprise (SALE) for external customers through AESIP.

A-15. **Global Combat Support System – Engineer (GCSS-EN).** The GCSS-EN is a tool used to support quantitative aspects of engineering support planning and execution. It provides the general requirements for the ESP and provides a common automated system for the joint force engineer planners to determine the appropriate amount of engineer assets and capabilities to support the selected COA. GCSS-EN is a web based application residing on the SECRET Internet Protocol Router Network (SIPRNET). GCSS-EN assists the engineer planners in determining the correct engineer capability for the proper location, timed correctly to support the concept of operations. GCSS-EN includes a TCMS module to assist with facilities planning and links into construction resource and materials planning. It also includes an environmental module. GCSS-EN is used to: generate time-phased facility requirements based on the OPLAN, analyze and assess engineering support by comparing facility requirements to in-theater facility -assets and HN, contract, and troop engineering capability, provide facility feasibility assessment, manpower, material, and nonunit cargo requirements for -other processes, provide infrastructure data to assist in mission analysis and COA development, and provide real time monitoring capability needed to track plan execution.

A-16. **Single Army Logistics Enterprise (SALE).** As described in the current Logistics Domain Strategic IT Plan, the Army has been making steady improvements over the last several years to develop a Single Army Logistics Enterprise (SALE), which will transition the Army from numerous independent and standalone IT systems to an integrated End-to-End (E2E) system operating in a net-centric environment. Over time, the SALE will be an integrated enterprise solution from the weapon system platforms to Army depots. The Army's overarching logistics architecture is the Army Integrated Logistics Architecture (AILA) which is guiding the Army logistics community in eliminating redundant IT investments and supports the integration and interoperability of the SALE in the Business Mission Area and Warfighter Mission Area. The Common Logistics Operating Environment (CLOE) is beginning to fuse logistics processes, embedded sensor-based information, and communication technologies to achieve a more interoperable, condition-based, and net-centric logistics enterprise that enable the SALE. The Army is establishing the Logistics Information Warehouse (LIW) to integrate data; fielding satellite communications to improve long distance connectivity; has flattened and consolidated tactical level logistics IT systems; is using radio frequency (RF) technology to track Army inventory, has mapped its ERP systems to the Defense

Logistics Management Standards (DLMS), is beginning to comply with the DOD-sponsored Item Unique Identification (IUID) marking system, and is making strides to establish an unprecedented capability that will enable equipment health management through a Condition-Base Maintenance Plus (CBM+) initiative. CBM+ is intended to increase operational readiness by repairing or replacing system components based on the actual condition of the component. These are all initiatives that either directly or indirectly help transition the Army's IT systems to the SALE.

A-17. **Battle Command Sustainment Support System (BCS3).** BCS3 is the logistics component of the Army Battle Command System (ABCS). Today, BCS3 is the Army's unclassified and classified logistics fusion center employed at multiple echelons for maneuver sustainment support. The system is for Army CSSBs, Stryker Brigades (BDE), and other BDEs/Groups/Regiments and distributed to HQ sections through brigade to theater.

A-18. **Force XXI Battle Command Brigade and Below (FBCB2).** FBCB2 is the principal digital C2 system for the U.S. Army at brigade level and below. The system is an automated, network enabled C2 system, which provides brigade and below elements with a seamless battle command capability. The system, positioned on specified platforms, will perform WFFs for the planning and execution of operations. FBCB2 is a component of Army Battle Command System (ABCS).

A-19. **Movement Tracking System (MTS).** The MTS is a vehicle based tracking and messaging system using commercial satellites (L-band), two-way free text messaging, digital maps, encryption, military Global Positioning System, and RFID interrogation. Using MTS Army sustainment organizations track in real time their truck locations, communicate with the drivers and redirect their cargo depending on the operational situation, and provide In Transit Visibility of those cargo assets containing RFID tags. This capability also provides the ability to avoid identified hazards, inform operators of unit location changes, and provide Traffic Regulation and Control. Common User Logistic Transport vehicles, selected combat and sustainment tactical wheeled vehicles, and watercraft can be fitted with MTS mobile units. MTS Control Stations are located in various combat support and sustainment C2 and staff sections to provide overview of multiple convoy operations within a specified area and to specifically control/direct those convoys specifically under their C2. In addition, a future interface between MTS and embedded equipment diagnostic and prognostic systems will provide accurate data that will aid fleet maintenance and improve availability and overall service life.

Standard Army Management Information System (Sustainment)

A-20. The current baseline of tactical sustainment Standard Army Management Information Systems (STAMIS) operate to support the war-fighter. These systems are fielded in Army logistics activities of the active and reserve components in virtually all TOE units and at the installation level in the Army. These systems provide modern automated logistics support for the functions of field maintenance, supply (inventory and materiel management), property accountability, ammunition, and readiness management. Through the use of new wireless data communications and portable satellite terminals, the tactical logistics STAMIS can communicate supply and maintenance transactions faster and over a greater distance to other STAMIS and to national providers. In addition, the introduction of Automatic Identification Technology (AIT) has brought increased asset visibility and source data automation capability that results in more accurate data input and faster response times from the tactical STAMIS.

A-21. **Property Book Unit Supply Enhanced**. The Property Book Unit Supply Enhanced is a web-based property accountability system that replaced the Standard Property Book System-Redesign and Unit Level Logistics Systems-S4. The system performs the functions of property accountability and unit supply required by AR 710-2 and AR 735-5 and DA Pamphlet 710-2-1,SR 735-30-1, Property Accountability (Supply and Property Accounting Procedures for Organizations and Units), SR 735-150-1, Property Accounting (Accounting for Lost, Damaged or Destroyed Property), AR 735-150-3 Property Accounting (Statement of Charges); and AR 735-150, Property Accounting (Accounting for Lost, Damaged; Destroyed Property. In tactical organizations, PBUSE systems will normally be located in: Company supply rooms, Property book offices, Battalion and Brigade S-4, and the G-4. PBUSE can operate in either the standalone or enterprise (connected to the Internet) mode. PBUSE capabilities include the maintaining of supported unit information, asset adjustments and lateral transfers, updating property book authorizations, and the production and management of unit hand receipts. It will generate and submit requests for Class II, Class IV, and Class VII equipment, follow-up and cancel supply transactions, post supply status, and receipt information to the activity register. It will also create and print hand receipts, property book, and activity register reports, manage basic and operational loads, request, receive, and turn-in ammunition.

A-22. **Unit Level Logistics System Aviation (ULLS-A) (E).** Company crew chiefs and unit level aviation maintenance personnel operate ULLS-A (E), a microcomputer based software system, to perform repair part supply and aviation maintenance management. It automates both supply chain management and the maintenance functions prescribed by The Army Maintenance Management System–Aviation, DA Pamphlet 738-751. The ULLS-A (E) system provides an enhanced aviation maintenance management capability. It is a multi-user system incorporating a Local Area Network (LAN) to link the functions of Tech Supply, Production Control and Quality Control, phase team, and back shop sections within the aviation field maintenance organization. The program incorporates a back shops module that gives the maintenance units the capability to initiate and complete work orders. The program also provides decision support and ad hoc query tools. This STAMIS will be integrated into the new enterprise solution, Global Combat Support System - Army (GCSS-A).

A-23. **Standard Army Maintenance System**. SAMS-E consists of both SAMS-1E and SAMS-2E applications and supports sustainment Table of Organization and Equipment unit level maintenance elements and Field and Sustainment maintenance shop production activities. For maneuver units, the SAMS-E systems are located at company and separate company level and can be consolidated at the battalion level; for combat support and sustainment elements. The SAMS-E systems provide consolidated maintenance and repair parts data. They are located at the Field Maintenance Teams, Forward Support Companies (FSC), and Maneuver Battalions, Field Maintenance Companies within the Maneuver BSBs, Separate Battalions and Brigades, and Sust Bdes, Sustainment Base Component Repair Companies (CRC), SMC, and Modular Force materiel management organizations. SAMS-E incorporates Windows graphical user interface operating systems (Windows XP). The system merges unit and field maintenance to act as a bridge between current functionality and the Enterprise Resource Planning (ERP) solution.

A-24. **Standard Army Retail Supply System (SARSS).** SARSS supports receipt, storage, issue, and management of Class II, Class IIIP, Class IV, and Class IX items of supply. SARSS is comprised of three interrelated subsystems SARSS-1, SARSS-2AC/B, and SARSS-Gateway. SARSS-1 is the standard supply system for receipts, storage, issues, replenishment, and storage operations. It is a real-time, transaction-oriented system where users can interactively enter, retrieve, and update supply information. SARSS-1 processes customer unit requests, cancellations, modifications, and follow-ups for supplies. SARSS-1 also provides an interactive query capability. The Materiel Release Order Capability and the Automated Manifest System are resident in the SARSS-1 baseline to control the flow of materiel, manage performance, and produce productivity reports. SARSS-2AC/B supports the Materiel Management requirements for all Class II, Class IIIP, Class IV, and Class IX processing. SARSS-2AC/B has asset visibility of SARSS-1 activities. Processes include all SARSS-2A functionality plus SARSS-2B non-time sensitive actions such as catalog, document history, demand history, and interface capability with financial systems. SARSS-Gateway offers improved communications and advanced automation functionality that allows users to place orders on the Source of Supply, the same day the customer produces them when not issued from on hand stocks.

A-25. **Standard Army Ammunition System Modernization (SAAS-MOD).** The SAAS-MOD is designed to provide centralized information management to support ammunition management functions on the battlefield and in garrison, within Army Commands, and ASCCs. As a multi-level automated ammunition management, reporting, and accounting system, SAAS-MOD automates all retail Class V management life-cycle functions. The system operates in both tactical and non-tactical environments and provides automation support for the TSC, ESC Distribution Management Centers, Ammunition Supply Activities at the Sust Bde and TSC levels (Theater Storage Areas and Ammunition Supply Points), Brigade Ammunition Office, and Ammunition Transfer Holding Points. The system uses AIT in receiving, shipping, and inventory procedures. SAAS-MOD will transition tactical ammunition functions to GCSS-Army at FOC. Garrison ammunition supply functions will transition to the Installation Fixed Base portion of Logistics Modernization Program at FOC.

A-26. **TC AIMS II.** TC AIMS II is the Army's unit deployment and theater operations (movement control) automated system. It provides critical planning data to JOPES and execution data to the Global Transportation Network. It facilitates provision of ITV and TAV using the RF AIT Server and tags. It provides planning data for units during deployment and visibility of cargo and passenger movement during execution. Provides an integrated information and data source for transportation activities during deployment, sustainment, and redeployment/retrograde operations. Complies with the Defense Transportation Regulation and employs DOD and Service shipment policies and procedures in peace and war for active and reserve forces. It automates origin shipping/receiving and deployment; sustainment and redeployment/retrograde processes; produces movement documentation and unit move data; and furnishes timely information. The system uses AIT in order to mark unit equipment for tracking during deployments. When the theater operations capability is used it automates the Transportation Movement Request, convoy clearance operations, and tasking of common use land transport.

A-27. **CMOS.** CMOS is being fielded to IMCOM and the Installation Transportation Office to support inbound and outbound freight operations. It will further the installation functions of ordering transportation, documenting the event, and providing for automated payments to commercial carriers. It will also be fielded to units that operate consolidated shipping and receiving points (CRSP) or similar cross dock operations in a deployed environment to facilitate receipt, sort, and ship functions in conjunction with the theater movement control procedures.

A-28. **Combat Service Support (CSS) Automated Information Systems Interface (CAISI).** Provides commercial and tactical network connections for sustainment Standard Army Management Information System (STAMIS), along with emerging systems. Users can be provided from a variety of locations in garrison or on the battlefield; ranging from the BCT and Brigade Support Areas to the fixed facilities within the tactical theater. The CAISI Client Module connects up to seven computers to wireless local area network (WLAN). The Bridge Module connects up to 14 computers to the WLAN. Data, voice, and media transmissions are extended beyond line-of-sight when CAISI is connected to sustainment Satellite Communications usually provided by sustainment Very Small Aperture Terminal (VSAT). The VSAT connectivity provides the CAIS WLAN users with access to the Non-secure Internet Protocol Router Network (NIPRNET) and Army/Defense Knowledge Online (AKO/DKO).

A-29. **The Installation Support Module Central Issue Facility (ISM CIF).** This system is required to manage personnel clothing issue records. It is a system that tracks all organizational Clothing and Individual Equipment (OCIE) issued by the Army to everyone in theater and allows the Army to ascertain the status of OCIE issued to units and individuals. All issues of OCIE from Central Issue Facilities (CIF) throughout the Army are captured on this system. During operations there will be fielding and exchange requirements of these items. Those transactions must be tracked on ISM CIF and supported in operations. All Soldiers in theater that have OCIE issued from a CIF in theater elsewhere must have a clothing record in ISM CIF and this must be updated as changes to those issues are made.

HEALTH MANAGEMENT SYSTEMS

A-30. Medical Communications for Combat Casualty Care (MC4) is the Army's medical information system. As the Army component of the deployed Defense Health Information Management System (DHIMS), MC4 will provide the hardware infrastructure for the DHIMS medical functionality software, as well as software required to ensure MC4/DHIMS interoperability with Army C2 and sustainment systems and provide reachback to the sustainment base. The MC4/DHIMS systems will rely on Army communications systems for transmission of health care information.

A-31. The DHIMS applications are built upon the functional, technical, and operational perspectives of theater and clinical medical professionals and will provide electronic health record in theater. DHIMS provides an integrated suite of software to support the military's deployed medical business practice. The theater family of systems supports complete clinical care documentation, medical supply and equipment tracking, patient movement visibility, and health surveillance in austere communications environments. A description of the theater DHIMS systems are described below.

- Armed Forces Health Longitudinal Technology Applications Mobile (AHLTA-Mobile). AHLTA Mobile is the first responder's handheld data capture device. It allows for immediate documentation of injury, illness, and care and stores medical data until it is transferred to AHLTA Theater. AHLTA Mobile can electronically store medical reference documents and replaces pounds of books and paper previously carried by medics.
- AHLTA Theater. AHLTA Theater extends the sustaining-base electronic medical record capability, look, and feel operation. It enables health care providers to document care; order laboratory services such as blood work, x-rays, and medications; and store medical data until communications are available to send the data to the Theater Medical Data Store and Clinical Data Repository.
- Theater Medical Information Program Composite Health Care System Cache (TC2). This system provides documentation for inpatient health care and ancillary services order-entry and result-reporting in the deployed environment. It provides inpatient management, laboratory, radiology, and pharmacy ordering capabilities. It also enables users to schedule outpatient clinic and radiological procedures.
- Theater Medical Data Store (TMDS). Information from the theater medical systems are transferred to the TMDS which serves as the authoritative theater database for collecting, distributing, and viewing Service members' pertinent medical information. TMDS updates the AHLTA Clinical Data Repository, where all Service members' electronic health records reside. This information is also made available to the VA through an interface known as Bidirectional Health Information Exchange–Theater. TMDS integrates the Joint Patient Tracking Application functionality to view, track, and disposition of ill or injured patients as they move through the theater levels of care, sustaining-base Military Treatment Facilities and those shared with the VA.
- Joint Medical Workstation (JMeWS). JMeWS provides medical situational awareness, medical surveillance, and force health decision support. It also reports on medical trends and analyzes the overall status of theater health. JMeWS provides the ability to drill down to specific medical units and individual encounters. It also shares intelligence with Global Combat Support System and Global Command and Control Systems, serving as the medical component to the Combatant and Joint Task Force Commander's common operating picture.
- DOD Occupational and Environmental Health Readiness System–Industrial Hygiene (DOEHRS-IH). DOEHRS-IH supports the reduction of worksite hazards and tracking of long-term environmental exposure. It provides analytical support for documenting occupational hazards by capturing analysis results of air, water, and soil samples.
- Defense Medical Logistics Standard System Customer Assistance Module (DCAM). DCAM is the medical logistics ordering tools that allows operational units to order and monitor Class VIII medical supplies and replenish levels when required. It automates the medical materiel supply process at lower levels of care and allows non-logisticians to electronically view and order from their supplier's catalog.

A-32. **Theater Enterprise-Wide Logistics System (TEWLS).** The TEWLS application is designed to transfer the capability for theater-level Class VIII supply chain management from TAMMIS into a Systems Applications and Products (SAP)-based enterprise architecture. The TEWLS AIS will build on the SAP ERP implementation started at the U.S. Army Medical Materiel Agency in May 2002 and would bring theater Class VIII management into the same system architecture that is used for the production of Army medical equipment sets and medical materiel sets. The TEWLS AIS supports the intermediate MEDLOG functions for distribution and materiel management and ties together the national, regional, and deployed units into a single business environment. It will support the development, production, and ultimate theater sustainment of medical assemblages that are the basic building blocks of operational medical capabilities. The TEWLS AIS will also support the operation of all Army organizations serving as the theater lead agent for medical materiel and provide materiel management within a single operational instance for tactical Army MLCs. Upon completion, the TEWLS application will migrate as an Army-sponsored initiative into the DMLSS program as the DMLSS theater-level solution for medical supply chain management.

Appendix B

NATO Standardization Agreements for Sustainment

This appendix is a reference for multinational Standardization Agreements (STANAGs) impacting on the sustainment of forces. Force developers should reference the applicable STANAG to ensure interoperability is achieved. Details regarding these STANAGs may be reviewed at http:\\nsa.nato.int.

1. NATO Logistics Handbook
2. AJP-4 - Allied Joint Logistics Doctrine
3. AJP-4.4 – Allied Joint Movement & Transportation Doctrine
4. AJP-4.5 – Allied Joint Host nation Support Doctrine and Procedures
5. AJP-3.4.1 – Peace Civil Support Operations
6. ALP-4.2 (STANAG 2406) – Land Forces Logistics Doctrine
7. APP-14 (STANAG 2284) – Land Compendium of Hand Arm Signals
8. AMovP-1 – Road Movements and Movements Control
9. AMovP-2 – Procedures for Movements Across National Frontiers
10. AMovP-3 – Movements and Transport Documents and Glossary of Terms and Definitions
11. AMovP-4 – Technical Aspects of the Transport of Military Materials by Railroad
12. AMovP-5 – Multi-Modal Movement and Transport Matters
13. AAP-4 – NATO Standardization Agreements and Allied Publications
14. AAP-6 – NATO Glossary of Terms and Definitions
15. AAP-15 – NATO Glossary of Abbreviations Used in NATO Documents and Publications
16. AAP-23 – NATO Glossary of Packing Terms and Definitions
17. AAP-24 (STANAG 2998) – Materials Handling Glossary of Terms and Definitions

18. STANAG 2034 – NATO Standard Procedures for Mutual Logistics Assistance
19. STANAG 2109 – Postal Organization and Courier Service of NATO Forces
20. AEP–13 (STANAG 2375) - Battlefield Vehicle Recovery & Evacuation Guide
21. STANAG 2399 - Battlefield Recovery/Evacuation Operation
22. AEP – 17 (STANAG 2400) – Battlefield Vehicle Recovery – User Handbook
23. STANAG 2413 - Demountable Load Carrying Platforms
24. STANAG 2418 - Policy for Expedient Repair, including Battle Damage Repair
25. STANAG 2494 – NATO Asset Tracking Shipping Label and Associated Symbology
26. STANAG 2827 - Materials Handling In The Field
27. STANAG 2828 – Military Pallets, Packages and Containers
28. STANAG 2830 - Materials Handling Aids
29. STANAG 2897 – Standardization of EOD Equipment Requirements and Equipment
30. STANAG 2926 – Procedures for the Use and Handling of Freight Containers for Military Supplies
31. STANAG 2927 – Marking of Restraint Equipment for Road Movement
32. STANAG 2961 - Classes of Supply of NATO Land Forces
33. STANAG 4062 – Slinging and Tie-Down Facilities for Lifting and Tying Down Military Equipment Movement by Land and Sea
34. STANAG 4280 – NATO Levels of Packing Note: Not ratified by US
35. STANAG 4281 – NATO Standard marking for Shipping and Storage
36. STANAG 4406 – Military Message Handling System

Glossary

This glossary lists acronyms and terms with Army, multi-Service, or joint definitions, and other selected terms. Where the Army and joint definitions are different, *(Army)* follows the term. Terms for which FM 4-0 is the proponent manual (the authority) are marked with an asterisk (*). The proponent manual for the other terms is listed in parentheses after the definition

AAFES	Army and Air Force Exchange Service
AAP	Allied administrative publication (NATO)
ABCA	American, British, Canadian, Australian (and New Zealand) Armies Program
ABCS	Army Battle Command System
ABS	automated battlebook system
ACOM	Army command
ACSA	acquisition and cross-servicing agreement
A/DACG	arrival/departure airfield control group
ADCON	administrative control
ADES	airdrop equipment and systems
ADS	aerial delivery support
AEP	Allied engineering publication (NATO)
AERS	airdrop equipment repair and supply
AFSB	Army field support brigade
AHLTA	Armed Forces Health Longitudinal Technology Application
AHS	Army Health System
AIT	automatic identification technology
AIS	automated information system
AJP	Allied joint publication (NATO)
ALP	Allied logistics publication (NATO)
ALT	acquisition, logistics, technology
AMC	Air Mobility Command
AMCOM	aviation and missile command
AMEDD	Army Medical Department
AMovP	Allied movement publication (NATO)
AO	area of operations
AOR	area of responsibility
APA	Army pre-positioned afloat
APOD	aerial port of debarkation
APOE	aerial port of embarkation
APP	Army procedural publication (NATO)
APS	Army pre-positioned stocks
AR	Army regulation

ARDEC	Army Research Development and Engineering Center
ARFORGEN	Army force generation
ARSOF	Army Special Operations Forces
ASA(ALT)	Assistant Secretary of the Army (Acquisition, Logistics, and Technology)
ASA(FM&C)	Assistant Secretary of the Army (Financial Management and Comptroller)
ASB	aviation support battalion
ASC	Army Sustainment Command
ASCC	Army Service component command
BCS3	Battle Command Sustainment Supply System
BCT	brigade combat team
bde	brigade
BH	behavioral health
BLS	brigade legal section
BOLT	brigade operational legal team
BOS	battlefield operating system
BSB	brigade support battalion
BSI	base support installation
C2	command and control
CAB	combat aviation brigade
CAISI	Combat Service Support Automated Information Systems Interface
CAPS	Commercial Accounts Payable System
CBRN	chemical, biological, radiological, nuclear
CCDR	combatant commander
CCIR	commander's critical information requirement
CE	communications-electronics
CJFLCC	combined joint force land component command
CLT	casualty liaison team
CMOS	Cargo Movement Operating System
CONUS	continental United States
COP	common operational picture
CRC	component repair company
CSA	Chief of Staff of the Army
CSB	contracting support brigade
CSH	combat support hospital
CSSB	combat sustainment support battalion
CUL	common user logistics
CVS	commercial vendor services
DA	Department of the Army
DAFL	directive authority for logistics

DA PAM	Department of the Army pamphlet
DBCAS	Database Commitment Accounting System
DCAM	Defense Medical Logistics Standard Support customer assistance module
DCIPS	Defense Casualty Information Processing System
DDS	Deployable Dispersing System
DFAS	Defense Financing and Accounting Service
DHIMS	Defense Health Information Management System
DIMHRS	Defense Integrated Military Human Resource System
div	division
DJMPS	Defense Joint Military Pay System
DLA	Defense Logistics Agency
DMC	distribution management center
DMLSS	Defense Medical Logistics Standard Support
DOD	Department of Defense
DODD	Department of Defense directive
DOM	Director of Materiel
DPO	distribution process owner
DS	direct support
DSCA	defense support of civil authroities
DTAS	defense theater accountability software
EA	executive agent
EAB	echelons above brigade
ECC	Expeditionary Contracting Command
EEFI	essential element of friendly information
ENCOORD	engineer coordinator
EOD	explosive ordnance disposal
EPS	essential personnel services
EPW	enemy prisoner of war
ERP	enterprise resource planning
ESC	expeditionary sustainment command
ESF	emergency support function
FBCB2	Force XXI Battle Command, Brigade and Below
FEMA	Federal Emergency Management Agency
FFE	field force engineering
FHP	force health protection
FM	field manual; financial management
FMC	financial management center
FMI	field manual, interim
FMP	field maintenance platoon
FMTP	financial management tactical platform

FORSCOM	United States Army Forces Command
FP	force provider
FRAGO	fragmentary order
FSC	forward support company
G-1	assistant chief of staff, personnel
G-4	assistant chief of staff, logistics
G-8	assistant chief of staff, financial management
GCC	geographic combatant commander
GCCS	Global Combat Support System
GCM	global container manager
GFEBS	General Funds Enterprise Business System
HBCT	heavy brigade combat team
HN	host nation
HNS	host nation support
HQ	headquarters
HR	human resources
HRS	human resources support
HRSC	human resources sustainment center
HSS	health service support
IA	implementing agreement
IATS	Integrated Automated Travel System
IGO	intergovernmental organization
IMCOM	Installation Management Command
IPB	information preparation of the battlefield
I/R	internment/resettlement
ISSA	inter-Service support agreement
ISB	intermediate staging base
ISO	International Organization for Standardization
ITV	in-transit visibility
J-1	personnel directorate of a joint staff
J-4	logistics directorate of a joint staff
JAGC	Judge Advocate General's Corps
JDDOC	joint deployment distribution operations center
JDOMS	Joint Director of Military Support
JFC	joint force commander
JFLCC	joint force land component command
JIIM	joint, interagency, intergovernmental, multinational
JM&L	joint munitions and lethality
JMC	Joint Munitions Command
JOA	joint operations area

JP	joint publication
JTF	joint task force
JTF-PO	Joint Task Force – Port Opening
LCMC	life cycle management command
LNO	liaison officer
LOC	line of communications
LOGCAP	logistics civilian augmentation program
LOTS	logistics over-the-short
MA	mortuary affairs
MC4	medical communications for combat casualty care
MCB	movement control battalion
MCD	Military Committee decision (NATO)
MCP	main command post
MCT	movement control team
ME	Materiel Enterprise
MEDBDE	medical brigade
MEDCOM	United States Army Medical Command
MEDCOM(DS)	medical command (deployment support)
MEDLOG	medical logistics
METT-TC	Mission, enemy, terrain and weather, troops and support available, time available and civil considerations
MHE	materials handling equipment
MICC	Mission and Installation Contracting Command
MILU	multinational integrated logistics unit
MLC	medical logistics company
MLMC	medical logistics management center
MMB	multifunctional medical battalion
MNF	multinational force
MOG	maximum aircraft on the ground
MRE	meal, ready to eat
MSA	mutual support agreement
MTF	medical treatment facility
MTS	Movement Tracking System
MWR	morale, welfare, and recreation
NAMSA	NATO Maintenance and Supply Agency
NATO	North Atlantic Treaty Organization
NEO	noncombatant evacuation operation
NGO	nongovernmental organization
NIPRNET	Nonsecure Internet Protocol Router Network
NRF	national response framework
NSE	national support element

OCONUS	outside the continental United States
OCP	operational command post
OE	operational environment
OGA	other government agency
OPCON	operational control
OPLAN	operations plan
OPLAW	operational law
OPORD	operation order
OPLOG	operations logistics
PA	personnel accounting
PARC	principal assistant responsible for contracting
PBUSE	property book unit supply enhanced
PCA	Posse Comitatus Act
PCC OTC	Paper Check Conversion Over The Counter
PEO	program (project) executive officer
PIM	personnel information management
PMESII-PT	political, military, economic, social, information, infrastructure, physical environment, time
POD	port of departure
POE	port of embarkation
PRM	personnel readiness management
PSA	port support activity
PSDR	personnel services delivery redesign
RDECOM	United States Army Research, Development, and Engineering Command
RF	radio frequency
RFID	radio frequency identification
RI	relevant information
RM	resource management
ROC	rehearsal of concept
ROE	rules of engagement
RS	religious support
RPMA	real property maintenance activities
RSOI	reception, staging, onward movement, integration
RTD	return to duty
S-1	personnel staff officer
S-4	logistics staff officer
SAAS-MOD	Standard Army Ammunition System - Modernized
SALE	Single Army Logistics Enterprise
SAMS	Standard Army Maintenance System
SAP	systems applications and products

SARSS	Standard Army retail Supply System
SBSO	Sustainment Brigade (Special Operations)
SCOE	Sustainment Center of Excellence
SDDC	Surface Deployment and Distribution Command
SECARMY	Secretary of the Army
SECDEF	Secretary of Defense
SJA	staff judge advocate
SMC	support maintenance company
SOFA	status-of-forces agreement
SOP	standing operating procedures
SPM	single port manager
SPO	support operations
SPOD	seaport of debarkation
SPOE	seaport of embarkation
SSA	supply support activity
STAMIS	standard Army management information system
STANAG	standardization agreement (NATO)
Sust Bde	sustainment brigade
SVC	stored value card
TA	Theater Army
TACOM	tank automotive and armaments command
TACON	tactical control
TAMMIS	the Army medical management information system
TAMMS-A	the Army maintenance management system - aviation
TC-AIMS II	Transportation Coordinator's Automated Information for Movement System II
TEWLS	theater enterprise-wide logistics system
TO	theater opening
TOE	table of organization and equipment
TPLSS	third-party logistics support services
TRADOC	United States Army Training and Doctrine Command
TSC	theater sustainment command
UCMJ	Uniform Code of Military Justice
ULLS-A	unit level logistics system - aviation
UMT	unit ministry team
UN	United Nations
USACC	United States Army Contracting Command
USACMA	United States Army chemical Materials Agency
USAFINCOM	United States Army Finance Command
USAHRC	United States Army Human Resources Command
USAMC	United States Army Materiel Command

USAMEDCOM	United States Army Medical Command
USARNORTH	United States Army, Northern Command
USASAC	United States Army Security and Assistance Command
USC	United States Code
USNORTHCOM	United States Northern Command
USTRANSCOM	United States transportation Command
VSAT	very small aperture terminal
WARNO	warning order
WFF	Warfighting Function
WLAN	wireless local area network
WRSA	war reserve stocks for allies

SECTION II – TERMS

alliance
(joint) The relationship that results from a formal agreement (such as a treaty) between two or more nations for broad, long-term objectives that further the common interests of the members (see JP 3-0).

***anticipation**
The ability to foresee events and requirements and initiate necessary actions that most appropriately satisfy a response.

Army Field Support Brigade (AFSB)
An organization which provides integrated and synchronized acquisition logistics and technology (ALT) support, less medical, to Army operational forces. (FM 4-93.41)

battle command
The art and science of understanding, visualizing, describing, directing, leading, and assessing forces to impose the commander's will on a hostile, thinking, and adaptive enemy. Battle command applies leadership to translate decisions into actions—by synchronizing forces and warfighting functions in time, space, and purpose—to accomplish missions. (FM 3-0).

Coalition
(joint) An ad hoc arrangement between two or more nations for common action (JP 5-0).

***continuity**
The uninterrupted provision of sustainment across all levels of war.

control
In the context of command and control, the regulation of forces and warfighting functions to accomplish the mission in accordance with the commander's intent. (FM 3-0)

Combat Sustainment Support Battalion (CSSB)
A flexible and responsive unit that executes logistics throughout the depth of an area of operations including transportation, maintenance, ammunition, supply, MA, airdrop, field services, water, and petroleum.(FM 4-93.2)

defensive operations
Combat operations that defeat an enemy attack, gain time, economize forces, and develop conditions favorable for offensive or stability operations (FM 3-0).

Directive Authority for Logistics
The Combatant Commander authority to issue directives to subordinate commanders. (Title 10 USC)

***distribution**
The operational process of synchronizing all elements of the logistics system to deliver the right things to the right place and right time to support the CCDR. It is a diverse process incorporating distribution management and asset visibility.

***distribution management**
The function of synchronizing and coordinating a complex of networks (physical, communications, information, and resources) and the sustainment WFF (logistics, personnel services, and HSS) to achieve responsive support to operational requirements.

***economy**
The provision of sustainment resources in an efficient manner to enable a commander to employ all assets to generate the greatest effect possible.

Essential elements of friendly information (EEFI).
A critical aspect of a friendly operation that, if known by the enemy, would subsequently compromise, lead to failure, or limit success of the operation, and therefore should be protected from enemy detection.

***field services**
The maintenance of combat strength of the force by providing for its basic needs and promoting its health, welfare, morale, and endurance.

general engineering
(joint)Those engineering capabilities and activities, other than combat engineering, that modify, maintain, or protect the physical environment. Examples include: the construction, repair, maintenance, and operation of infrastructure, facilities, lines of communication and bases, and terrain modification and repair and selected explosive hazard activities (JP 3-34).

generating force
Those Army organizations whose primary mission is to generate and sustain the operational Army's capabilities for employment by JFCs. (FM 1-01)

***Human Resource Sustainment Center**
A multifunctional, modular organization (staff element) assigned to a TSC that provides HR support to the theater.

health services support
All support and services performed, provided, and arranged by the AMEDD to promote, improve, conserve, or restore the mental and physical well being of personnel in the Army and, as directed in other Services, agencies and organizations. (FM 4-02.12)

host nation support
(joint) Civil and military assistance rendered by a nation to foreign forces within its territory during peacetime, crises or emergencies, or war based on agreements mutually concluded between nations. (JP 1-02)

***integration**
The joining all of the elements of sustainment (tasks, functions, systems, processes, and organizations) to operations assuring unity of purpose and effort.

intelligence preparation of the battlefield (IPB)
A continuous staff planning activity undertaken by the entire staff to understand the operational environment and options it presents to friendly and threat forces. (FM 2-0)

interagency coordination
The coordination that occurs between elements of Department of Defense and engaged U.S. Government agencies for the purpose of achieving an objective (FM 3-0).

Internment/Resettlement operations
To take or keep selected individuals in custody or control as a result of military operations to control their movement, restrict their activity, provide safety, and/or gain intelligence (FM 3-19.40).

***In-theater reconstitution**
The extraordinary actions that commanders take to restore a degraded unit to combat effectiveness commensurate with mission requirements and available resources.

information system
(Army) Equipment and facilities that collect, process, store, display, and disseminate information. This includes computers—hardware and software—and communications, as well as policies and procedures for their use. (FM 3-0)

***improvisation**
The ability to adapt sustainment operations to unexpected situations or circumstances affecting a mission.

liaison
(joint) Contact or intercommunication maintained between elements of military forces or other agencies to ensure mutual understanding and unity of purpose and action (JP 3-08).

legal support
The provision of professional legal services at all echelons. (FM 1-04)

Logistics
(joint) The planning and executing the movement and support of forces. It includes those
aspects of military operations that deal with: design and development, acquisition, storage, movement, distribution, maintenance, evacuation, and disposition of materiel; movement, evacuation, and hospitalization of personnel; acquisition or construction, maintenance, operation, and disposition of facilities; and acquisition or furnishing of services (JP 4-0).

***maintenance**
All actions taken to retain materiel in a serviceable condition or to restore it to serviceability

multinational integrated logistics unit
(NATO) A unit which is formed when two or more nations agree, under OPCON of a NATO commander, to provide logistics support to a MNF. (ALP 4.2)

national support element
(NATO) Any national organization or activity that supports national forces that are a part of a Multi National Force (ALP 4.2)

offensive operations
Combat operations conducted to defeat and destroy enemy forces and seize terrain, resources, and population centers. They impose the commander's will on the enemy (FM 3-0)

operating forces
Those forces whose primary missions are to participate in combat and the integral supporting elements thereof (FM1-01).

operational environment
(joint) The composite of the conditions, circumstances, and influences that affect the employment of capabilities and bear on the decisions of the commander (JP 3-0).

operational contract support
(joint) The process of planning for and obtaining supplies, services, and construction from commercial sources in support of operations along with the associated contractor management functions. (JP 4-10)

***personnel services**
Those sustainment functions maintaining Soldier and Family readiness and fighting qualities of the Army force.

***regeneration**
The rebuilding of a unit. It requires large-scale replacement of personnel, equipment, and supplies.

relevant information
All information of importance to commanders and staffs in the exercise of command and control. (FM 3-0).

***reorganization**
Action to shift resources within a degraded unit to increase its combat effectiveness.

***responsiveness**
The ability to meet changing requirements on short notice and to rapidly sustain efforts to meet changing circumstances over time

stability operations
(joint) An overarching term encompassing various military missions, tasks, and activities conducted outside the United States in coordination with other instruments of national power to maintain or reestablish a safe and secure environment, provide essential governmental services, emergency infrastructure reconstruction, and humanitarian relief. (JP 3-0)

***supply**
The procurement, distribution, maintenance while in storage, and salvage of supplies, including the determination of kind and quantity of supplies.

***survivability**
The ability to protect personnel, information, infrastructure, and assets from destruction or degradation

sustainment
The provision of logistics, personnel services, and health services support (HSS) necessary to maintain operations until mission accomplishment (FM 3-0).

sustainment brigade (Sust Bde)
A flexible, multifunctional sustainment organization, tailored and task organized according to METT-TC. (FM 4-93.2)

***Sustainment maintenance**
Maintenance which is generally characterized as "off system" and "repair rear".

sustainment Warfighting Function (WFF)
The related tasks and systems that provide support and services to ensure freedom of action, extend operational reach, and prolong endurance (FM 3-0).

third party logistic support services
(NATO) The use of preplanned civilian contracting to perform selected sustainment. Its aim is to enable competent commercial partners to provide a proportion of deployed sustainment so that such support is assured for the commander and optimizes the most efficient and effective use of resources (ALP 4.2)

***transportation**
The moving and transferring of personnel, equipment, and supplies to support the concept of operations, including the associated planning, requesting, and monitoring.

Unified action
(joint) The synchronization, coordination, and/or integration of the activities of governmental and nongovernmental entities with military operations to achieve unity of effort. (JP 1)

References

SOURCES USED

These are the sources quoted or paraphrased in this publication. These sources contain relevant supplemental information.

JOINT AND DEPARTMENT OF DEFENSE PUBLICATIONS

Most joint publications are available online: http://www.dtic.mil/doctrine/jpcapstonepubs.htm.

DODD 5525.5. DoD *Cooperation with Civilian Law Enforcement Officials.* 15 January 1986.

JP 1. *Doctrine for the Armed Forces of the United States.* 14 May 2007.

JP 1-0. *Personnel Support to Joint Operations.* 16 October 2006.

JP 1-06. *Financial Management Support in Joint Operations.* 04 March 2008.

JP 3-0. *Joint Operations.* 17 September 2006.

JP 3-08. *Interagency, Intergovernmental Organization, and Nongovernmental Organization Coordination during Joint Operations Volumes I and II.* 17 March 2006.

JP 3-34. *Joint Engineer Operations.* 12 February 2007.

JP 3-35. *Deployment and Redeployment Operations.* 7 May 2007.

JP 4-0. *Joint Logistics. 18 July 2008.*

JP4-02. *Health Service Support.* 31 October 2006.

JP 4-06. *Mortuary Affairs in Joint Operations.* 5 June 2006.

JP 4-10. *Operational Contract Support.* 17 October 2008.

ARMY PUBLICATIONS

Most Army doctrinal publications are available online: https://akocomm.us.army.mil/usapa/doctrine/.

FM 1. *The Army.* 14 June 2005.

FM 1-0. *Human Resources Support.* 21 February 2007.

FM 1-01. *Generating Force Support for Operations.* 2 April 2008.

FM 1-04. *Legal Support to the Operational Army.* 15 April 2009.

FM 1-05. *Religious Support.* 18 April 2003.

FM 1-06. *Financial Management Operations.* 21 September 2006.

FM 3-0. *Operations.* 27 February 2008.

FM 3-05.140. *Army Special Operations Forces Logistics.* 12 February 2009.

FM 3-11.5. *Multiservice Tactics, Techniques, and Procedures for Chemical, Biological, Radiological, and Nuclear Decontamination.* 4 April 2006.

FM 3-28.1. *Multiservice Tactics, Techniques, and Procedures for Civil Support Operations.* 3 December 2007.

FM 3-34. *Engineer Operations.* 2 April 2009.

FM 3-34.400. *General Engineering.* 9 December 2008.

FM 3-35.1. *Army Pre-positioned Operations.* 1 July 2008.

FM 3-90. *Tactics.* 4 July 2001.

FM 4-01.30 *Movement Control.* 1 September 2003.

FM 4-02. *Force Health Protection in a Global Environment.* 13 February 2003.

FM 4-02.1. *Combat Health Logistics.* 28 September 2001.

FM 4-02.10. *Theater Hospitalization.* 3 January 2005.

FM 4-20.64. *Mortuary Affairs Operations.* 9 January 2007.

FM 4-30.1. *Munitions Distribution in the Theater of Operations.* 16 December 2003.

FM 4-30.13. Ammunition *Handbook: Tactics, Techniques, and Procedures for Munitions Handlers.* 1 March 2001

FM 4-30.3. *Maintenance Operations and Procedures.* 28 July 2004.

FM 4-93.4. *Theater Support Command.* 15 April 2003.

FM 5-0. *Army Planning and Orders Production.* 20 January 2005.

FM 6-0. *Mission Command: Command and Control of Army Forces.* 11 August 2003.

FM 6-22. *Army Leadership.* 12 October 2006.

FM 10-1. *Quartermaster Principles.* 11 August 1994.
FM 10-27. *General Supply in Theaters of Operation.* 20 April 1993.
FM 55-1. *Transportation Operations.* 3 October 1995.
FM 55-60. *Army Terminal Operations.* 15 April 1996.
FM 55-80. *Army Container Operations.* 13 August 1997.
FM 63-23. *Aviation Support Battalion.* 6 June 1996.
FM 100-10-1. *Theater Distribution.* 1 October 1999.
FM 100-10-2. *Contracting Support on the Battlefield.* 4 August 1999.
FM 100-15, *Corps Operations.* 29 October 1996.
FMI 1-0.01. *S-1 Operations.* 10 October 2007.
FMI 1-0.02. *Theater-Level Human Resources Operations Support.* 20 February 2008.
FMI 3-35. *Army Deployment and Redeployment.* 15 June 2007.
FMI 4-93.2. *The Sustainment Brigade.* 4 February 2009.
FMI 4-93.41. *Army Field Support Brigade Tactics, Techniques, and Procedures.* 22 February 2007.
FMI 5-0.1. *The Operations Process.* 31 March 2006.

DOCUMENTS NEEDED

These documents must be available to the intended users of this publication.

JP 1-02. *Department of Defense Dictionary of Military and Associated Terms.* 12 April 2001.

READINGS RECOMMENDED

These readings contain relevant supplemental information.

None

Index

Made in the USA
Monee, IL
04 November 2026

67382836R00066